AF581769

Waveform Design for 5G and beyond Systems

Waveform Design for 5G and beyond Systems

Editor

Kwonhue Choi

MDPI • Basel • Beijing • Wuhan • Barcelona • Belgrade • Manchester • Tokyo • Cluj • Tianjin

Editor
Kwonhue Choi
Department of Informaion and
Communication Engineering
Yeungnam University
Gyeongsan
Korea, South

Editorial Office
MDPI
St. Alban-Anlage 66
4052 Basel, Switzerland

This is a reprint of articles from the Special Issue published online in the open access journal *Electronics* (ISSN 2079-9292) (available at: www.mdpi.com/journal/electronics/special_issues/waveform_5g_systems).

For citation purposes, cite each article independently as indicated on the article page online and as indicated below:

LastName, A.A.; LastName, B.B.; LastName, C.C. Article Title. *Journal Name* **Year**, *Volume Number*, Page Range.

ISBN 978-3-0365-3175-5 (Hbk)
ISBN 978-3-0365-3174-8 (PDF)

Contents

About the Editor

Kwonhue Choi

Kwonhue Choi (Senior Member, IEEE) received B.S., M.S., and Ph.D. degrees in electronic and electrical engineering from the Pohang University of Science and Technology, Pohang, South Korea, in 1994, 1996, and 2000, respectively. From 2000 to 2003, he worked for the Electronics and Telecommunications Research Institute, Daejeon, South Korea, as a Senior Research Staff Member. In 2003, he joined the Department of Information and Communication Engineering, Yeungnam University, Gyeongsan, South Korea, where he is currently a professor. He has authored a textbook: *Problem-Based Learning in Communication Systems Using MATLAB and Simulink* (Wiley, 2016). His research interests include signal design for the communication systems, multiple access schemes, diversity schemes for wireless fading channels, multiple antenna systems, and in-band full duplex systems. He has invented advanced waveforms for wireless communications such as LP-FBMC. FRAC-OFDM, and FD-FBMC.

Preface to "Waveform Design for 5G and beyond Systems"

5G traffic has very diverse requirements with respect to data rate, delay, and reliability. The concept of using multiple OFDM numerologies adopted in the 5G NR standard will likely meet these multiple requirements to some extent. However, the traffic is radically accruing different characteristics and requirements when compared with the initial stage of 5G, which focused mainly on high-speed multimedia data applications. For instance, applications such as vehicular communications and robotics control require a highly reliable and ultra-low delay. In addition, various emerging M2M applications have sparse traffic with a small amount of data to be delivered. The state-of-the-art OFDM technique has some limitations when addressing the aforementioned requirements at the same time. Meanwhile, numerous waveform alternatives, such as FBMC, GFDM, and UFMC, have been explored. They also have their own pros and cons due to their intrinsic waveform properties. Hence, it is the opportune moment to come up with modification/variations/combinations to the aforementioned techniques or a new waveform design for 5G systems and beyond. The aim of this Special Issue is to provide the latest research and advances in the field of waveform design for 5G systems and beyond.

Kwonhue Choi
Editor

MDPI

Editorial

Waveform Design for 5G and beyond Systems

Kwonhue Choi 

Department of Information and Communication Engineering, Yeungnam University, Gyeongsan 712-749, Korea; gonew@yu.ac.kr

1. Aims and Objectives

Currently, 5G communication systems are being commercially deployed in many countries. It is also true that several new applications are expected to be realized within the pre-defined 5G capabilities. To accommodate these various new applications, 5G traffic has very diverse requirements for data rate, delay, and reliability. Such challenging requirements are met when the deployment of 5G is based on proper radio access technology. Consequently, multiple OFDM numerologies were adopted in the 5G NR (New Radio) standard, i.e., legacy OFDM of the 4G system was scaled up to cope with the requirements for enhanced mobile broadband, enhanced machine-type communications, and ultra-reliable low latency communications.

The concept of using multiple OFDM numerologies likely meets the multiple requirements for a 5G system to some extent. However, this cannot be an ultimate solution, because the traffic is radically accruing more varied characteristics and requirements when compared to the initial stage of 5G, which focused mainly on the high-speed multimedia data application. For instance, applications such as vehicular communications and robotic control require high reliability and ultra-low delay. In addition, various emerging machine-to-machine (M2M) applications have sparse traffic, with a small amount of data to be delivered. The state-of-the-art OFDM technique has some limitations when addressing the aforementioned requirements at the same time. Meanwhile, numerous waveform alternatives, such as FBMC, GFDM, and UFMC, have been explored to overcome this situation. However, they also have their pros and cons due to their intrinsic waveform properties. Hence, it is very timely to come up with modifications/variations/combinations to the waveform alternatives to OFDM, together with the state-of-the-art OFDM technique, or a new innovative waveform design for 5G systems and beyond. This Special Issue aims to provide the latest research and advances in the field of waveform design, along with the related issues of 5G systems and beyond.

Citation: Choi, K. Waveform Design for 5G and beyond Systems. *Electronics* **2021**, *10*, 2124. https://doi.org/10.3390/electronics10172124

Received: 6 August 2021
Accepted: 26 August 2021
Published: 1 September 2021

2. Review of the Contributions in This Issue

As previously mentioned, this Special Issue collects the latest research and advances in the field of waveform design for 5G systems and beyond, and issues related to it. This Special Issue presents five high-quality articles on these subjects. The contents of the articles in the Special Issue are summarized below.

It is important to note that the research on waveform design needs to come with an exact performance assessment and well-balanced comparison with their competitors. In light of this objective, the first pair of articles, [1,2], provide intensive comparisons among the various waveforms used for 5G systems and beyond, and the employed pulse shapes, respectively. In the first article "Field Trials of SC-FDMA, FBMC, and LP-FBMC in Indoor Sub-3.5 GHz Bands" [1], a common issue in conventional methodologies for the comparison among the promising waveforms for the 5G system and beyond was addressed. These conventional comparisons all have the same critical limitations that the results were based on—a stereotyped channel model and the simple nonlinearity model of analog circuits. They are substantially different from the performance results of the waveforms in a real channel with a real transceiver. To address this difference, the authors have compared the performances of three waveforms, i.e., SC-FDMA, FBMC, and LP-FBMC, in a real uplink indoor channel. From these studies, they figured out that the LP-FBMC is a suitable waveform for real indoor applications. The second article, "An Overview of FIR Filter Design in Future Multicarrier Communication Systems" [2], presents a comprehensive survey on the recent advances in finite impulse response (FIR) filter design methods in multicarrier modulation (MCM)-based communication systems. First, the fundamental aspects were presented, including an introduction to existing waveform candidates and the principle of FIR filter design. Then, the methods of FIR filter design were summarized, focusing on the following three categories: frequency sampling methods, windowing-based methods, and optimization-based methods. Finally, the performances of various FIR design methods were evaluated and quantified by power spectral density (PSD) and bit error rate (BER). This paper also discusses different MCM schemes, as well as their potential prototype filters.

Furthermore, the Special Issue also addresses another aspect of waveform design, considering the fact that multi-antenna transmission has become a basic and crucial element in communication systems. A lot of work is performed on multi-antenna signal design, in conjunction with the employed waveforms. Accordingly, this Special Issues includes two articles, [3,4], which are focused on advanced signal design for space-time multi-antenna techniques for 5G systems and beyond. In the article, "Waveform Design for Space-Time Coded MIMO Systems with High Secrecy Protection" [3], the authors have presented a novel secrecy-enhanced waveform design for the quasi-orthogonal space-time block code scheme. The proposed waveform embeds dynamic artificial interference (AI) that depends on the signal, as well as on the channel gain of the legitimate receiver. The dependency of the AI on the information signal is used to reduce the difference in power allocations across the transmit antennas. In addition, the dependency of the AI on the channel gain is used to cancel the AI only at the legitimate receiver, while imposing serious interference on the passive eavesdropper. As a result, the secrecy capacity of the waveform is highly enhanced. Since the proposed method does not incur any power loss due to the addition of AI, it can be efficiently utilized in many wireless systems using STBC, resulting in diversity gain, as well as security protection. As part of this category of research, the article "Improved Pair-Wise Detections of Differential Quasi-Orthogonal Space-Time Modulation with Four Transmit Antennas" [4] proposes a new complex and real pair-wise detection for conventional differential space-time modulations, based on a quasi-orthogonal design with four transmit antennas for general QAM. Owing to the independent joint ML detection of two complex and real symbol pairs, respectively, the decoding complexity is the same as, or lower than, the conventional differential detections. The proposed detections exhibit almost identical performance to an optimum maximum-likelihood receiver, as well as an improved performance when compared with conventional pair-wise detections, especially for higher modulation order.

Last but not least, the topic of this Special Issue is multiple access (MA) capability, which is one of the most crucial measures in designing waveforms. The last article "Performance Analysis of LDS Multi Access Technique and New 5G Waveforms for V2X Communication" [5] contributes to the waveform design for 5G and beyond, concerning multiple access capability. The authors of this article have proposed new efficient MA schemes, where the low-density signature (LDS) structure is combined with two waveforms for 5G systems and beyond, i.e., universal filtered multi-carrier (UFMC) and filtered-OFDM waveforms. They also have simulated their proposed schemes under a vehicular channel of a vehicle to everything (V2X) communication. From this simulation, it was shown that the LDS-F-OFDM significantly achieves higher performance improvements when compared with the LDS-OFDM and LDS-UFMC in all scenarios, while maintaining an affordable complexity at the transmitter and receiver sides. These improvements are attributed to the advantages that the filtered-OFDM waveform offers through addressing the adequate filter.

Funding: This work was partly supported by the Technology development Program of MSS [S2967489], the NRF grant funded by the Korea government (MSIT) (No. 2021R1A2C1010370), and the 2021 Yeungnam University Research Grant.

Acknowledgments: As the Guest Editor, I convey gratitude to all the authors of this Special Issue for their great contributions. All the peer reviewers who helped to evaluate the submitted manuscripts and made beneficial suggestions are also gratefully acknowledged. I also would like to thank the editorial board of *Electronics*, who graciously invited me to guest edit this Special Issue. Finally, I owe my special thanks to Xenia Xie at the *Electronics* Editorial Office. The timely completion of the Special Issue would not have been possible without her diligence.

Conflicts of Interest: The author declares no conflict of interest.

References

1. Na, D.; Jang, S.; Seo, W.-G.; Choi, K. Field Trials of SC-FDMA, FBMC and LP-FBMC in Indoor Sub-3.5 GHz Bands. *Electronics* **2021**, *10*, 573. [CrossRef]
2. Jiang, L.; Zhang, H.; Cheng, S.; Lv, H.; Li, P. An Overview of FIR Filter Design in Future Multicarrier Communication Systems. *Electronics* **2020**, *9*, 599. [CrossRef]
3. Shang, P.; Lee, H.; Kim, S. Waveform Design for Space–Time Coded MIMO Systems with High Secrecy Protection. *Electronics* **2020**, *9*, 2003. [CrossRef]
4. Kim, H.; Shang, Y.; Kim, S.; Jung, T. Improved Pair-Wise Detections of Differential Quasi-Orthogonal Space-Time Modulation with Four Transmit Antennas. *Electronics* **2021**, *10*, 1675. [CrossRef]
5. Kholouani, I.; Elbahhar, F.; Elassali, R.; Idboufker, N. Performance Analysis of LDS Multi Access Technique and New 5G Waveforms for V2X Communication. *Electronics* **2020**, *9*, 1094. [CrossRef]

Article

Field Trials of SC-FDMA, FBMC and LP-FBMC in Indoor Sub-3.5 GHz Bands

Dongjun Na [1], Sangmin Jang [1], Won-Gi Seo [2] and Kwonhue Choi [1,*]

1 Department of Information and Communication Engineering, Yeungnam University, Gyeongsan 38541, Korea; nadj2964@ynu.ac.kr (D.N.); smjang@ynu.ac.kr (S.J.)
2 Nextwill, Yuseong-gu, Daejeon 34155, Korea; nextwill@nextwill.com
* Correspondence: gonew@yu.ac.kr

Abstract: LP-FBMC (low peak-to-average power ratio filter bank multicarrier) was recently proposed to ameliorate the high peak-to-average power ratio (PAPR) issue of filter bank multicarrier (FBMC). The previous simulation study showed that LP-FBMC achieves a similar PAPR as that of single carrier frequency division multiple access (SC-FDMA) while being very robust to inter-user timing/frequency offsets. However, the simulation results that were obtained assuming the stereotyped channel model and the simple nonlinearity model of analog circuits substantially differ from the performance results in a real channel with a real transceiver. To address this, the main purpose of this work is to compare the performances of three waveforms, i.e., SC-FDMA, FBMC, and LP-FBMC, in a real uplink indoor channel. We investigate how the bit error rate (BER) performance gaps of three waveforms in the indoor channels change by the system parameters, such as the carrier frequency within sub-3.5 GHz band and the number of sub-carriers or the sub-carrier spacing, which was not found in the previous simulation study. Our investigation confirms that LP-FBMC is a suitable waveform for real indoor applications.

Keywords: 5G waveform; SC-FDMA; FBMC; Low PAPR FBMC (LP-FBMC); access timing offset; carrier frequency offset; high-power amplifier (HPA) nonlinearity; software defined radio (SDR) device; uplink indoor channel; out-of-band (OOB) emission

Citation: Na, D.; Jang, S.; Seo, W.-G.; Choi, K. Field Trials of SC-FDMA, FBMC and LP-FBMC in Indoor Sub-3.5 GHz Bands. *Electronics* **2021**, *10*, 573. https://doi.org/10.3390/electronics10050573

Academic Editor: Paolo Colantonio

Received: 18 December 2020
Accepted: 27 January 2021
Published: 1 March 2021

1. Introduction

The single-carrier frequency division multiple access (SC-FDMA) scheme was adopted as the uplink scheme in 4G and 5G wireless networks [1]. SC-FDMA has lots of merits, such as low implementation complexity, low peak-to-average power ratio (PAPR), and ease of application to massive multiple-input and multiple-output (MIMO) systems. However, the SC-FDMA system is very sensitive to inter-user timing/frequency offsets and it has high out-of-band (OOB) emission, and solving these problems has been challenging [2,3]. Especially, for the heterogeneous network or the massively connected applications in 5G network scenario, the waveforms with highly localized spectrum are required to mitigate interference between users (or systems) in adjacent bands caused by inter-user (or inter-system) timing/frequency offsets. Thus, many studies on new multiple access schemes are being conducted in order to prepare the standards for the future generations [4–6].

As an alternative to SC-FDMA for next-generation networks, various modulation waveforms are being considered, such as the filtered orthogonal frequency division multiplexing (f-OFDM), filter bank multicarrier (FBMC), universal filtered multicarrier (UFMC), and generalized frequency division multiplexing (GFDM) schemes. Among those waveforms, FBMC has the lowest OOB emission and, thus, inter-user interference that is caused by inter-user timing/frequency offsets can be eliminated using only one guard subcarrier. Additionally, inter-carrier interference that is caused by a multipath fading channel can be overcome without using a cyclic prefix (CP) [6,7]. Because of these benefits, FBMC has been considered to be a promising candidate for multiple access schemes of next-generation communications. Meanwhile, an enhanced version of FBMC, called LP-FBMC (low PAPR FBMC), was proposed to reduce the PAPR of the conventional FBMC by applying discrete Fourier transform (DFT) spreading with special conditions [8–10].

The performances of SC-FDMA, conventional FBMC, and LP-FBMC were intensively compared by computational simulations in [6]. It was shown that LP-FBMC achieves a substantially improved bit error rate (BER) performance than SC-FDMA in a uplink channel with inter-user timing/frequency offsets and signal clipping. However, it is difficult to generalize the simulation results in a theoretical channel model to the real channel performance. This is because too many irregular factors to simulate are involved in the real channel response. For instance, the multipath fading and shadow fading are highly subject to the physical propagation environments, and the hardware impairment factors, such as transceiver's digital-to-analog-converter/analog-to-digital-converter nonlinearities, oscillators' phase noises, and high-power amplifier (HPA) nonlinearity, are determined by the radio frequency (RF) circuitry of the transceiver and the carrier frequency band. In addition, the problem with the artificial computational experimental model for channels or hardware impairments is that it is likely to be biased or friendly to one particular waveform. To address this, the main purpose of this work is to compare these three waveforms, i.e., SC-FDMA, FBMC, and LP-FBMC, in a real uplink indoor channel from various perspectives and confirm that LP-FBMC is a practically competitive solution for beyond-5G indoor applications.

In the 5G standard, the millimeter wave (mmWave) band of 28 GHz and the mid-band below 6 GHz are used interchangeably, and the mid-band range is allocated for indoor wireless networks, because mmWave signals suffer severe propagation attenuation in indoor environments [11]. Various studies on the SC-FDMA system in indoor wireless communication bands are being widely conducted, because SC-FDMA has been adopted as the standard uplink access scheme since 4G [12–15]. On the other hand, there has not been a field test study on FBMC (or modified version of FBMC) in indoor wireless communication bands, and it has not been verified that FBMC (or modified version of FBMC) can solve the problems of SC-FDMA in the real indoor wireless environment. For reference, there have been few studies on the spectrum shape of FBMC waveforms [16,17].

In this study, we intensively analyse the performance of SC-FDMA, FBMC, and LP-FBMC via a field test in a real indoor wireless environment. To this end, we implement a test bed utilising a software defined radio (SDR) device. As for the indoor wireless channel, we test the various sub-3.5 GHz bands when considering the different indoor wireless band allocations in the different countries (3.5 GHz band for South Korea; 600 MHz, 2.5 GHz, and 3.5 GHz bands for US; and, 700 MHz and 3.5 GHz bands for Europe [18]). Specifically, the three waveforms (The three uplink schemes in comparison, SC-FDMA, FBMC, and LP-FBMC differ basically in the waveforms. Hence, the terms *'three waveforms'* and *'three uplink schemes'* are used interchangeably for referring to {SC-FDMA, FBMC, LP-FBMC} in this paper.) are compared in the following aspects:

- spectral density of the transmit waveform affected by the RF circuit and the HPA;
- point-to-point link performance according to distance and carrier frequency in two distinct environments: line-of-sight (LoS) and non-line-of-sight (NLoS) environments;
- uplink BER performance according to the time/frequency offset among the users, distance, and carrier frequency; and,
- uplink BER performance according to the subcarrier spacing, and the number of subcarriers.

The remainder of this paper is structured, as follows. Section 2 describes the mathematical waveform models of SC-FDMA, FBMC, and LP-FBMC. Section 3 describes the SDR test-bed for the field experiments in indoor wireless channels. The BER performances are intensively tested and compared among the waveforms from various perspectives in Section 4. Finally, we make conclusions in Section 5.

2. Waveform Models for SC-FDMA, FBMC, and LP-FBMC

2.1. SC-FDMA

Figure 1 shows the structure of an SC-FDMA transmitter for the kth uplink user. The number of users is denoted by K, the number of carriers allocated to each user is denoted by N, and the number of symbols per data frame is denoted by M. The m-th complex data symbol transmitted to the n-th subcarrier under SC-FDMA is expressed as $x_{n,m}^{(k)}$, and the m-th complex data symbol vector is expressed as $\mathbf{x_m^{(k)}} = [x_{0,m}^k, x_{1,m}^k, \ldots, x_{N-1,m}^k]^T$, where $[\cdot]^T$ denotes the transposed matrix. In Figure 1, $\mathbf{X_m^{(k)}}$ is the N-point discrete Fourier transform (DFT) output for input $\mathbf{x_m^{(k)}}$. If the n-th component of $\mathbf{X_m^{(k)}}$ is expressed as $X_{n,m}^{(k)}$, then the k-th user's transmit signal $s^{(k)}(t)$ can be expressed, as follows [6]:

$$s^{(k)}(t) = \sum_{m=0}^{M-1}\sum_{n=0}^{N-1} p(t - mT_s)X_{n,m}^{(k)}e^{j\frac{2\pi\alpha(n)}{T}(t-mT_s)}, \quad (1)$$

where T is the data symbol duration, T_s is the symbol duration, including the cyclic prefix (CP), and $p(t) = \begin{cases} 1 \text{ for } 0 \le t < T_s \\ 0 \text{ elsewhere} \end{cases}$. The variable $\alpha(n)$ is the subcarrier index that is allocated to the nth element of N-point DFT output of the k-th user. We consider the block sub-band (SB) allocation, and one guard subcarrier between adjacent subcarrier blocks, which are common to FBMC and LP-FBMC for fair comparison [19]. Therefore, $\alpha(n)$ is expressed as $\alpha(n) = k(N+1) + n$.

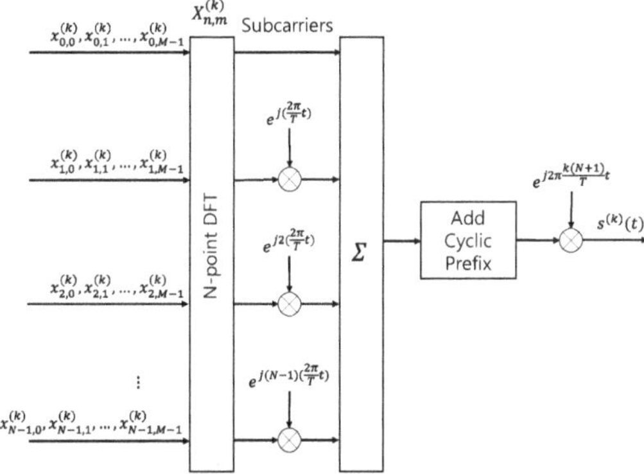

Figure 1. Structure of the k-th user's single carrier frequency division multiple access (SC-FDMA) transmitter.

2.2. FBMC

Figure 2 shows the structure of a conventional FBMC transmitter for the k-th uplink user's, where $R\{\cdot\}$ and $I\{\cdot\}$ are the real and imaginary parts of the input, and $h(t)$ represents the impulse response of a prototype filter [20]. The m-th complex data symbol transmitted to the n-th subcarrier of the k-th user denoted by $x_{n,m}^{(k)}$ is expressed, as follows:

$$x_{n,m}^{(k)} = a_{n,m}^{(k)} + jb_{n,m}^{(k)}, 0 \le n \le N-1,\ 0 \le m \le M-1, \quad (2)$$

where $a_{n,m}^{(k)}$ and $b_{n,m}^{(k)}$ are the real and imaginary parts of the $x_{n,m}^{(k)}$. Subsequently, the FBMC signal of k-th user $s^{(k)}(t)$ is expressed, as follows [8]:

$$s^{(k)}(t) = \sum_{n=0}^{N-1}\sum_{m=0}^{M-1}\left\{ a_{n,m}^{(k)}h(t-mT) + jb_{n,m}^{(k)}h(t-mT-\tfrac{T}{2}) \right\}e^{j\frac{2\pi\alpha(n)}{T}t}e^{jn\frac{\pi}{2}}, \quad (3)$$

where $\alpha(n)$ is expressed in the same way as in SC-FDMA.

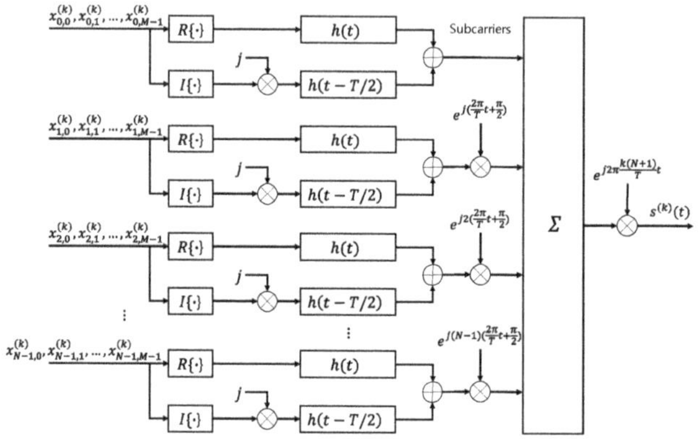

Figure 2. Structure of the k-th user's filter bank multicarrier (FBMC) transmitter.

2.3. LP-FBMC

In [8], it was reported that simply applying DFT spreading to FBMC cannot achieve the PAPR reduction effect, and there exist a special condition of phase coefficients to maximize the single carrier effect of DFT spreading for PAPR reduction. This condition is called the identically time-shifted multicarrier (ITSM) condition. In addition, in order to further reduce the PAPR, LP-FBMC generates four ITSM-conditioned signal candidates for each subframe and concatenates the ones with the lowest PAPR to form a frame. Figure 3 shows the transmitter structure for the k-th subframe of LP-FBMC. The four versions of signal candidates are generated according to two switching control bits S_1 and S_2, as shown in Table II of [8].

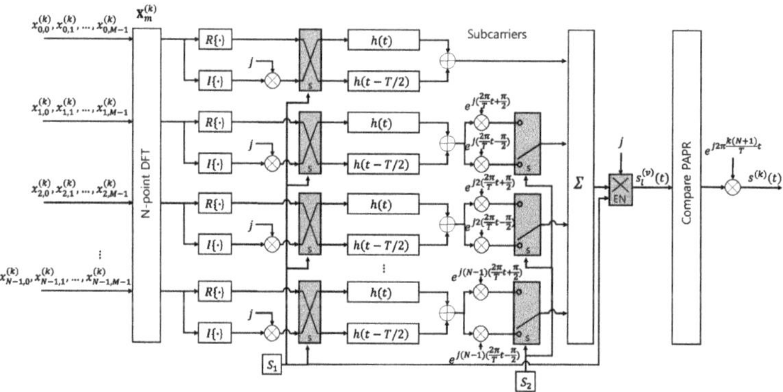

Figure 3. Structure of the the k-th user's low peak-to-average power ratio filter bank multicarrier (LP-FBMC) transmitter.

In Figure 3, a complex vector $\mathbf{X}_{\mathbf{m}}^{(\mathbf{k})} = [X_{0,m}^{k}, X_{1,m}^{k}, \cdots, X_{N-1,m}^{k}]^{T}$ denotes the DFT output vector of complex data symbols, and the n-th element of $\mathbf{X}_{\mathbf{m}}^{(\mathbf{k})}$ is given, as follows:

$$X_{n,m}^{(k)} = A_{n,m}^{(k)} + jB_{n,m}^{(k)}, 0 \leq n \leq N-1,\ 0 \leq m \leq M-1, \tag{4}$$

where $A_{n,m}^{(k)}$ and $B_{n,m}^{(k)}$ denote the real and imaginary parts of $X_{n,m}^{(k)}$, respectively. The transmitter structures of LP-FBMC and FBMC are almost the same, except that the DFT is performed on the modulated data symbols, and two switches were added, as shown in Figures 2 and 3. After generating four transmission signal candidates according to the two switches, the candidate signal with the lowest PAPR is selected for transmission. In the l-th data block of the k-th user, the v-th candidate of transmit signal $s_l^{(k)(v)}(t)$, $l = 0, 1, 2, \ldots, M/W - 1$, can be expressed, as follows [10]:

$$s_l^{(k)(v)}(t) = \sum_{n=0}^{N-1} \sum_{m=lW}^{W(l+1)-1} \left\{ A_{n,m}^{(k)} p_{n,m}^{(v)}(t) + B_{n,m}^{(k)} q_{n,m}^{(v)}(t) \right\} e^{j\frac{2\pi\alpha(n)}{T}t}, \; v = 1,2,3,4, \tag{5}$$

where W denotes the number of offset quadrature amplitude modulation (OQAM) symbols in the l-th data block, and $p_{n,m}^{(v)}(t)$ and $q_{n,m}^{(v)}(t)$ are expressed, as follows [10]:

$$p_{n,m}^{(v)}(t) = \begin{cases} h(t - mT)e^{jn\left(\frac{2\pi}{T}t + \frac{\pi}{2}\right)} & \text{if } v = 1 \\ h(t - mT)e^{jn\left(\frac{2\pi}{T}t - \frac{\pi}{2}\right)} & \text{if } v = 2 \\ jh\left(t - \frac{T}{2} - mT\right)e^{jn\left(\frac{2\pi}{T}t + \frac{\pi}{2}\right)} & \text{if } v = 3 \\ jh\left(t - \frac{T}{2} - mT\right)e^{jn\left(\frac{2\pi}{T}t - \frac{\pi}{2}\right)} & \text{if } v = 4, \end{cases} \tag{6}$$

$$q_{n,m}^{(v)}(t) = \begin{cases} jh\left(t - \frac{T}{2} - mT\right)e^{jn\left(\frac{2\pi}{T}t + \frac{\pi}{2}\right)} & \text{if } v = 1 \\ jh\left(t - \frac{T}{2} - mT\right)e^{jn\left(\frac{2\pi}{T}t - \frac{\pi}{2}\right)} & \text{if } v = 2 \\ -h(t - mT)e^{jn\left(\frac{2\pi}{T}t + \frac{\pi}{2}\right)} & \text{if } v = 3 \\ -h(t - mT)e^{jn\left(\frac{2\pi}{T}t - \frac{\pi}{2}\right)} & \text{if } v = 4. \end{cases} \tag{7}$$

At the final stage, we successively concatenate the selected candidates. The concatenated waveform up to the l-th block of the k-th user $c_l^{(k)}(t)$ can be expressed, as follows [10]:

$$c_l^{(k)}(t) = c_{l-1}(t) + s_l^{(k)(u_l^k)}(t), \text{ with } c_{-1}(t) = 0, \tag{8}$$

where u_l^k is a candidate index that is selected in the l-th data block of the k-th user.

3. Test-Bed and Field Trial Environment

3.1. Test-Bed Configuration

In order to implement a test-bed in a real wireless channel, we used an SDR device (USRP-2901) and LabVIEW Comms 3.0 as a software platform. Table 1 shows the RF specifications of the SDR device, and Figure 4 shows the overall configuration of the test-bed. A Vert900 antenna was used for performance tests in the sub-1 GHz band, and an LP0965 antenna was used in the band higher than 1 GHz.

The overall operation processes of the test-bed are as follows. On a host PC, the transmit signal is modulated using LabVIEW Comms 3.0 and it is sent to the SDR device. In the SDR device, the modulated signal is converted into a continuous analog signal and it passes through I-Q mixing and up-conversion. Subsequently, an RF signal is generated and it is propagated to the chosen frequency band through the RF antennas, Vert-900 or LP0965. The SDR device of the receiver performs the reverse process of the transmitter. The received signal is transmitted to the host PC, and then the host PC performs synchronization, demodulation, and bit detection.

Table 1. RF specifications of the software defined radio (SDR) device and the antennas.

	Item	Value
SDR device: USRP-2901	Frequency range	70 MHz to 6 GHz
	Maximum output power	20 dBm
	Maximum input power	−15 dBm
	Maximum instantaneous real-time bandwidth	56 MHz
Antenna	LP0965 antenna gain	5–6 dBi
	Vert900 antenna gain	3 dBi

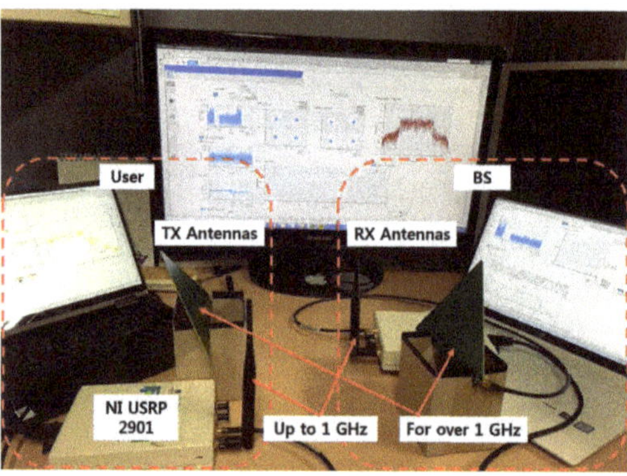

Figure 4. Test-bed configuration.

3.2. Indoor Wireless Channel Environment Set-Up

Line-of-sight (LoS) and non-line-of-sight (NLoS) environments were both implemented to measure the BER performance in typical indoor situations. The BER measurement was performed in the various spots on the 3rd floor in the IT building at Yeungnam University, Korea.

Figures 5 and 6 shows the transmitter (TX) (or user) and receiver (RX) (or base station (BS)) location set-up for LoS and NLoS environments, respectively, in the floor map of the building. The BS was tested at each of the three different positions specified with BSi ($i = 1, 2, 3$). In the uplink system, the inter-user interference (IUI) from the very next (left and right) sub-band (SB) is dominant over the those from the other SBs. Hence, for an efficiently down-scaled field test, three consecutive subbands are allocated to three uplink users (See Figure 7), and, then, we measured the performance of only the middle user who undergoes the IUIs from the left and right SBs.

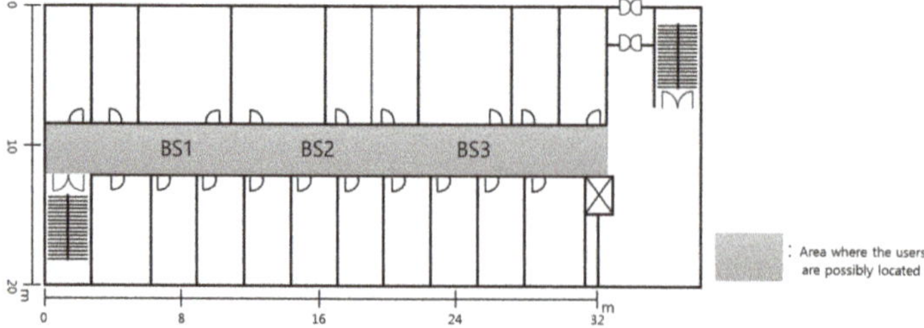

Figure 5. User area (shaded) and base station (BS) location in line-of-sight (LoS) environment.

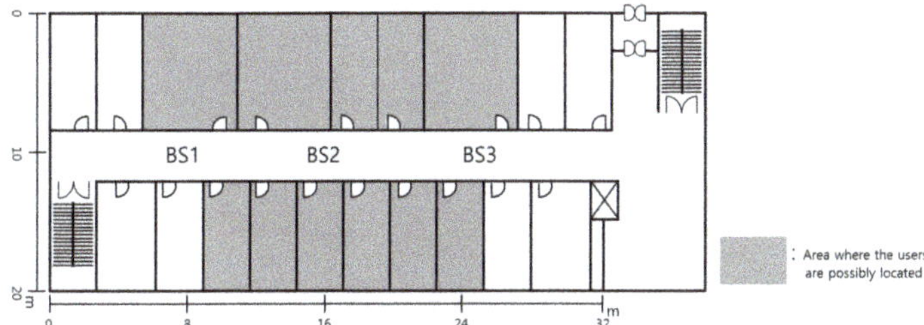

Figure 6. User area (shaded) and BS location in non-line-of-sight (NLoS) environment.

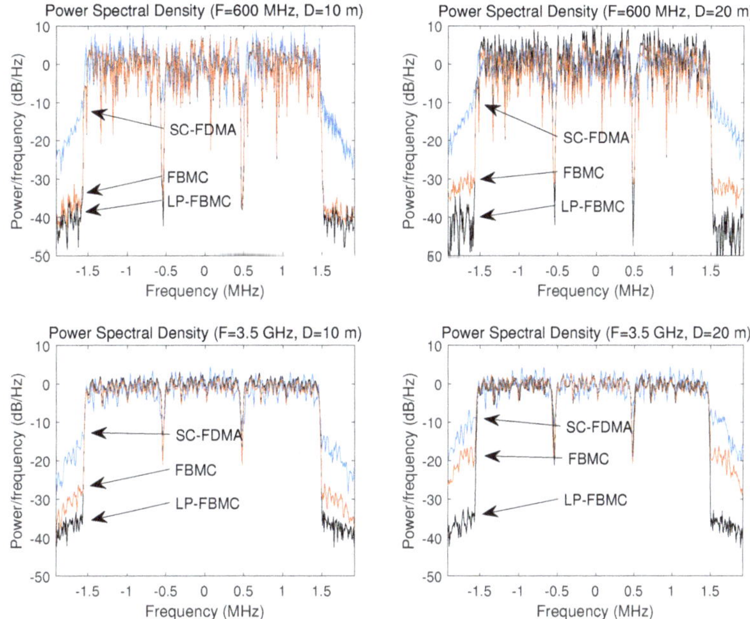

Figure 7. Power spectra of the SC-FDMA , FBMC, and LP-FBMC, $K = 3$, $N = 64$, LoS environment, $\Delta f = 15$ kHz.

The maximum straight distance between the TX and BS is 24 m and 12 m for LoS and NLoS environments, respectively. In the NLoS environment, the TXs were placed inside the rooms (shaded area in Figure 6), and the BS was placed in the corridor. Thus, there are more than one wall between TX and BS. Figure 8 is a sample shot of the NLoS environment set-up. The locations for the three equipments labelled by BS1, BS2, and BS3 in Figure 8 correspond to the points that are specified by BS1, BS2, and BS3 in Figures 5 and 6, respectively. At each of three BS positions, the average uplink BER was separately measured and then it was averaged. Note that, by the channel reciprocity of either direction between two positions, i.e., TX $\rightarrow$ BS, and TX $\leftarrow$ BS, the field test results that are obtained with the TX and BS location combinations in Figures 5 and 6 inherently include the case when the locations of TX and BS are switched.

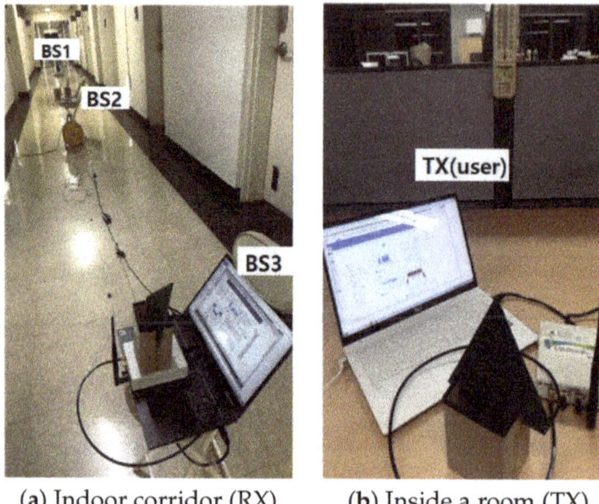

(**a**) Indoor corridor (RX) (**b**) Inside a room (TX)

Figure 8. Bit error rate (BER) performance measurement set-up for NLoS environment.

To synchronize the transmission(access) timings of the uplink users, or to apply the intended (To realize the effect of the practical inter-user synchronization process, we intentionally apply timing offsets at the transmission stage. The field test results for this case are coverer in Section 4.3.) timing offsets among the users, the SDR devices of the users' TXs are wired to a clock distribution device. Figure 9 shows the configuration of host PCs for users, SDR devices for users' TXs, and the clock distribution device.

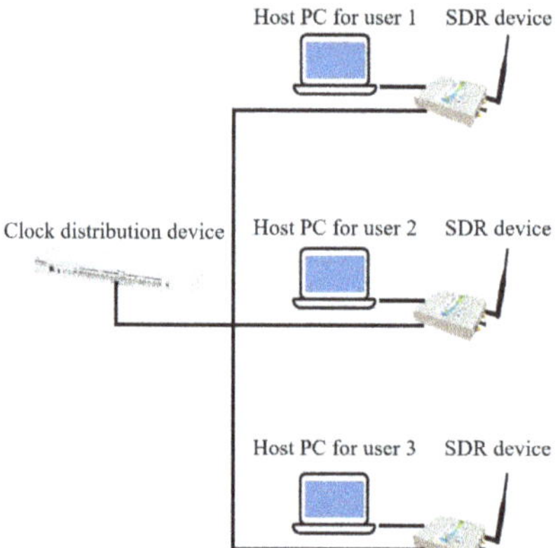

Figure 9. Configuration of host PCs for users, SDR devices for users' transmitters (TXs), and clock distribution device.

4. Experiment Results

In this section, we analyse and compare the temporal and spectral characteristics of the three waveforms, i.e., SC-FDMA, FBMC, and LP-FBMC, and then we compare the field-tested BER performance of the three waveforms in various real uplink wireless channels. Table 2 shows the major system parameters that are used for the performance measurements.

Table 2. Major system parameters.

Parameters	Value
Carrier frequency F	0.6 GHz (USA), 0.7 GHz (Europe), 2.5 GHz (USA), 3.5 GHz (USA, Europe, Korea)
Number of users K	3
Number of sub-carriers per user N	64, 128
Sub-carrier spacing $\Delta f(=1/T)$	15 kHz, 30 kHz
Oversampling factor	2
Modulation scheme	QPSK
Number of symbols per frame M	20
Filter for FBMC	PHYDYAS pulse with overlapping factor = 4 [20]
Cyclic prefix for SC-FDMA	T/4

4.1. Spectra of SC-FDMA, FBMC, and LP-FBMC

In a previous study [6], we confirmed that LP-FBMC is robust to signal clipping, because it achieves low PAPR. However, because a HPA in RF circuitry has different nonlinearities, the actual transmit signal may be significantly different from the result assuming a simple nonlinear model, such as signal clipping. In addition, the theoretical PAPR curve has a limitation for quantitatively assessing how the waveform degrades with this practical nonlinearity. Hence, we compare the spectra of SC-FDMA, FBMC, and LP-FBMC when they go through the actual nonlinearity of RF circuits in the SDR device.

Figure 7 shows the received power spectra of the three waveforms when the carrier frequency is 600 MHz and 3.5 GHz in the LoS environment of Figure 5, and the uplink power control is performed so that the power of each user is the same in the BS. The number of subcarriers N is 64, the subcarrier spacing Δf is 15 kHz, and the distances between the transmitter and receiver are 10 m and 20 m. In accordance with the well known drawback of OFDM-based waveforms, it is shown that SC-FDMA has the highest OOB emission among the three waveforms. Whereas, since the FBMC-based waveforms are filtered for each carrier, the its OOB emission is quite suppressed as compared to SC-FBMC. However, the field test results shown in Figure 7 also confirmed that the OOB emission of FBMC increases significantly as the distance increases from 10 m to 20 m. This is because, as the distance increases, the transmit power is accordingly increased to compensate path loss and, thus, the FBMC undergoes signal distortion due to its high PAPR characteristics. This issue is even more critical in millimeter band (i.e., 3.5 GHz in our field test), where the path loss is more severe when compared to the lower frequency bands (i.e., 600 MHz in our field test). On the other hand, the OOB emission of the LP-FBMC is maintained almost the same, irrespective the frequency band and the distances. Based on these results, the LP-FBMC system maintains the superb OOB suppression, not only in theoretical signal clipping model, but also in real RF circuits and wireless channels.

It is also shown in Figure 7 that the indoor wireless channel in the 600 MHz band has higher frequency selectivity than in the 3.5 GHz band. This confirms a well known property that the multipath delay spread gets smaller as the frequency increases.

4.2. BER Comparison According to Carrier Frequency and Distance

Figure 10 shows the BER performance of the three waveforms, according to the distance between RX and TX (=D) and carrier frequency (=F). For the averaged performance over the area, 10 different locations in the shaded areas in Figures 5 and 6 for LoS and NLoS environments, respectively, were tested for each pair of (D, F), and then the measured BERs are averaged. In general, as D increases, the TX power amplification gain (power gain in short) should be increased in order to compensate the path loss. However, if the power gain is set too high considering only the pass loss, then the performance is degraded by the nonlinear distortions of the transceiver circuit. Figure 11 shows the received signal power versus the power gain of the employed SDR device (USRP-2901) for different distances. In order to see the overall end-to-end link nonlinearities, i.e., not only HPA in TX, but also LNA (low noise amplifier), down-conversion circuit, and analog-to-digital converter (ADC) in RX, the received signal power was measured at the ADC output of RX. Meanwhile, TX and RX were placed at the outdoor open space to exclude the channel effect in the nonlinearity measurement. The received power tends to saturate when the power gain is larger than a certain level (about 18 dB) by the nonlinear distortion, as shown in Figure 11. Except that, due to the path loss, the graphs shift vertically according to the distance, the overall nonlinear characteristics remain the same, irrespective of the distance. Note that, even with the smaller received power due to the increased distance (12 m), the received power still saturates at the identical range of power gain. This implies that the main factor for the overall non-linearity is in the TX side, i.e., HPA in the TX. Hence, we experimentally searched the optimal power gain at each location of the user. Figure 12 shows an example plot of the measured BER versus the power gain for the three waveforms in LoS environment when F is 700 MHz and distance between TX and RX is 12 m. It is shown that the BER is minimized at a gain of 16 dB for FBMC and at a gain of 17 dB for LP-FBMC and SC-FDMA. The smaller optimal power gain for FBMC when compared to those for LP-FBMC and SC-FDMA confirms the higher PAPR of FBMC as compared to those for LP-FBMC and SC-FDMA. This is because, as PAPR becomes larger, the power gain needs to be more tightly controlled to avoid nonlinear distortion.

In Figure 10, commonly to the three waveforms, the BER performance deteriorates as the distance increases. This is because, as the distance increases, even optimal transmit power gain cannot fully compensate the path loss in order to avoid the unacceptable nonlinear distortion of the HPA. Hence, in the NLoS environment where the path loss is more serious as compared to LoS environment, BER degrades more rapidly as the distance increases. The BER degradations are more critical in higher carrier frequency due to higher path loss.

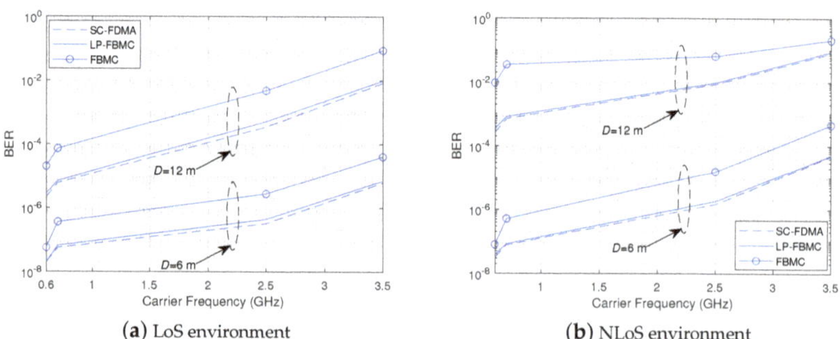

(**a**) LoS environment (**b**) NLoS environment

Figure 10. BER of the SC-FDMA, FBMC, and LP-FBMC, according to the distance and carrier frequency, $N = 64$, $\Delta f = 15$ kHz.

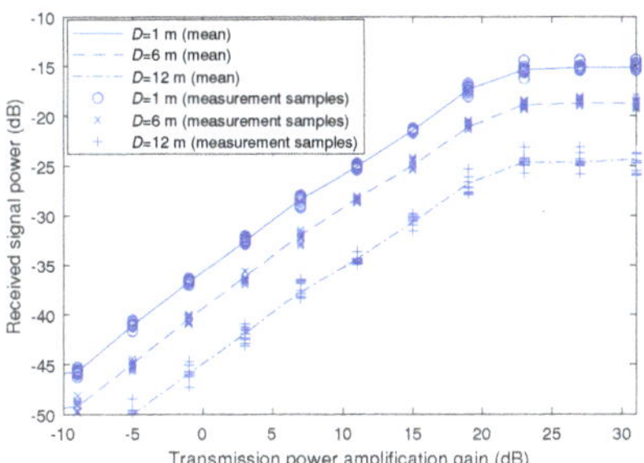

Figure 11. Received signal power versus the transmission power amplification gain of the SDR device (USRP-2901) in an outdoor open space.

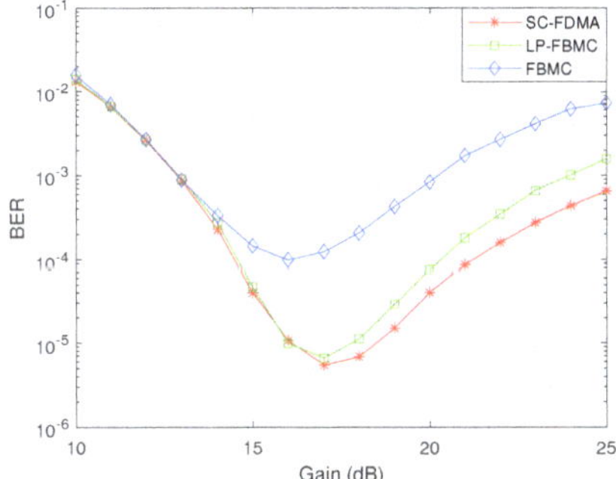

Figure 12. BER of the SC-FDMA, FBMC, and LP-FBMC according to the transmit power amplification gain, $F = 700$ MHz, distance between TX and RX = 12 m, LoS environment.

FBMC has the significantly higher BER when compared to SC-FDMA due to high PAPR. The BER gap between FBMC and SC-FDMA increases as the carrier frequency decreases with larger distance. This is because. as the carrier frequency decreases, the frequency selectivity increases and SC-FDMA exploits frequency diversity by DFT spreading, whereas FBMC cannot exploit frequency diversity. Hence, in the NLoS environment where the channel is more frequency-selective when compared to LoS environment, the BER gap between FBMC and SC-FDMA further increases. On the other hand, LP-FBMC achieves almost the same BER as that of SC-FDMA over the entire considered region of distance and carrier frequency. This is because LP-FBMC has similar PAPR performance as that of SC-FDMA, and its signal generation also includes DFT spreading. All of these trends are common to LoS and NLoS environments.

4.3. BER Comparison According to Timing/Frequency Offsets

Even with inter-user (inter-node) synchronization, there exist inevitable timing and frequency offsets among the uplink users. This is due to the imperfect dynamics of ranging process between BS and the users or inexpensive oscillators having insufficient resolution in the low cost user equipments. Note that the global reference timing set-up in Figure 8 provides perfect synchronization among the users' TXs. Hence, in our field test, we apply intentional timing/frequency offsets to the modulated waveforms for various offset ranges among the users. This allows for us to realize the practical inter-user synchronisation effects without being tackled with the implementation of the complicated inter-user synchronisation process. Note that the term *'timing/frequency offsets'* here refers to the offsets among the users, i.e., inter-user offsets. These offsets cannot be eliminated by the basic receiver synchronization process, because the receiver (BS) synchronization process only separately tracks the frequency and timing of each user's received signal.

In addition, a power control process was performed in order to ensure that the received powers of the user signals are identical. To this end, we first optimize the transmit power amplification gain of the furthest user from the BS by trading off between high signal strength and low signal distortion in HPA. Subsequently, we adjust the amplification gain of the other users, so that the received signal power is the same as the furthest user.

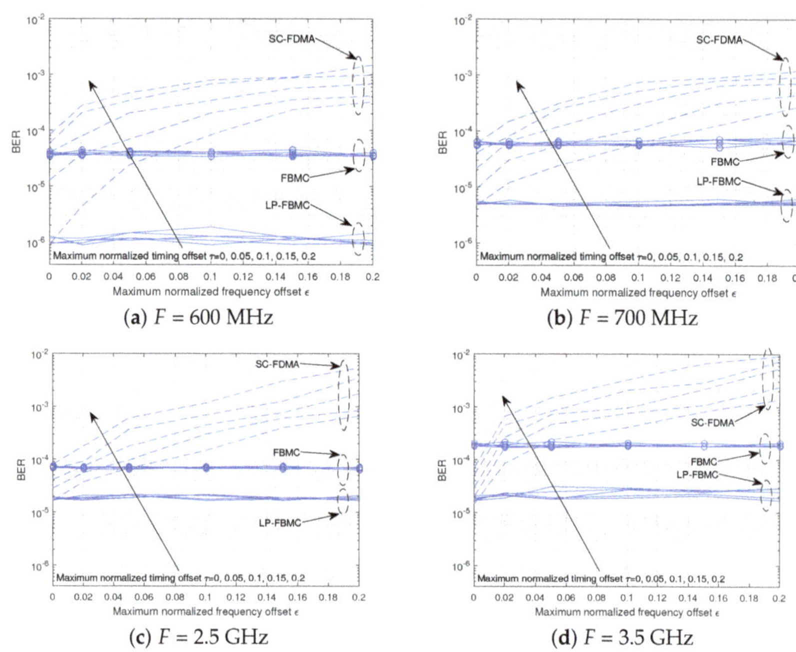

Figure 13. BER of the SC-FDMA, FBMC, and LP-FBMC, according to timing/frequency offset for various carrier frequencies, $N = 64$, $\Delta f = 15$ kHz.

Figure 13 shows the field-tested uplink BER results for various carrier frequencies for the case when each uplink user's timing and frequency offsets are uniformly distributed in $\left[-\frac{\tau T}{2}\ \frac{\tau T}{2}\right]$ and in $\left[-\frac{\epsilon}{2T}\ \frac{\epsilon}{2T}\right]$, respectively. The parameter τ denotes the symbol interval (=T)-normalized inter-user timing offset limit and the parameter ϵ denotes the subcarrier spacing (=1/T)-normalized inter-user frequency offset limit. Hence, τ and ϵ are unitless parameters. We randomly generated timing/frequency offset for each data frame until the average measured BER converges to obtain the averaged performance over the various uplink users' link environment. In addition, 60 different combinations of the users' locations randomly distributed in all of the shaded areas in Figures 5 and 6 were tested. The measurement procedure is structured, as follows: first, once the three user positions have been set, we perform out the aforementioned power control process. Second, the BERs were separately measured for all pairs of (τ,ϵ) in Figure 13. Third, for each of 60 combinations of the users' locations, the first and second steps that are mentioned above are repeated. Lastly, 60 measured BERs for each pair of (τ,ϵ) are averaged. Note that, in order to fully include inter-user interference, we measured the BER of the user allocated to the middle of the three consecutive subbands, as mentioned previously.

It is shown that the BERs of SC-FDMA abrubtly increase as τ and ϵ increase. Even with low PAPR advantage of SC-FDMA over FBMC, SC-FDMA performs worse than FBMC if τ and ϵ become larger than the corresponding limits. These agree with computational simulation results shown in [6]. However, note that, as compared to the theoretical simulation results of [6] (see Figures 5, 6 in [6]), the performance degradation of SC-FDMA by inter-user interference is more drastic in Figure 13. This comes from the difference between the channel that is assumed in the computational simulation of [6] and the actual channels in our field test. In [6], the Vehicular-A channel model in the international telecommunications union-radio communication (ITU-R) sector was used. Thus, it is revealed that there is more inter-user interference in the real indoor channel than in the Vehicular-A channel. Note that this trend is most emphasized in the lowest carrier frequency, i.e., F = 600 MHz in our field test. This implies that the inter-user interference becomes more detrimental as the frequency selectivity increases. Meanwhile, the BERs of FBMC and LP-FBMC are almost invariant to τ and ϵ. In particular, no matter how large τ and ϵ are given in the considered region, LP-FBMC achieves the identical BER performance to that of perfectly synchronized SC-FDMA. Again, the BER performance gains of LP-FBMC over FBMC and SC-FDMA are biggest in the lowest carrier frequency, i.e., F = 600 MHz.

4.4. BER Comparison According to Per-User Bandwidth

We can change the bandwidth (equivalently data rate) per user by two different ways. One is to change the subcarrier spacing while keeping the number of users and the other way is to change the number of subcarriers while keeping the subcarrier spacing. Figure 14 shows the BER results with the same system parameters as those for Figure 13d, except that the per-user bandwidth is enlarged twice as that of Figure 13d by doubling the number of sub-carriers N. It is shown that the BER of LP-FBMC decreases due to increased frequency diversity by increased DFT-spreading size (equal to N). This also holds for SC-FDMA, but it is limited to the case with perfect inter-user synchronization. Whereas, owing to lack of DFT-spreading, FBMC is relatively much worse than SC-FDMA and LP-FBMC.

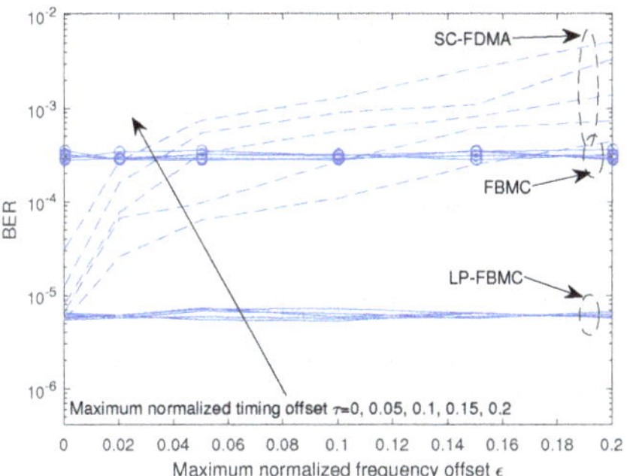

Figure 14. BER of the SC-FDMA, FBMC, and LP-FBMC according to timing/frequency offset, $N = 128$, Δf = 15 kHz, F = 3.5 GHz.

In 5G and beyond, various subcarrier spacings (i.e., $\Delta f (=\frac{1}{T})$ = 15 kHz, 30 kHz, and 60 kHz) are used to support high-quality service [1]. Figure 15 shows the case when the per-user bandwidth is enlarged twice by doubling the subcarrier spacing. For doubling the subcarrier spacing, the symbol duration decreases to one half of that presented in Figure 13d and, thus, the CP length ($=T/4$) also decreases to one half. The reduced CP results in larger inter-symbol interference (ISI) to SC-FDMA. This explains a remarkable result in Figure 15 that SC-FDMA more severely deteriorates with increasing timing offset as compared to the case in Figure 14. Summing up, the BER performance gain of LP-FBMC over SC-FDMA is more increased for lower carrier frequency, larger subcarrier spacing, and larger timing/frequency offset among the users.

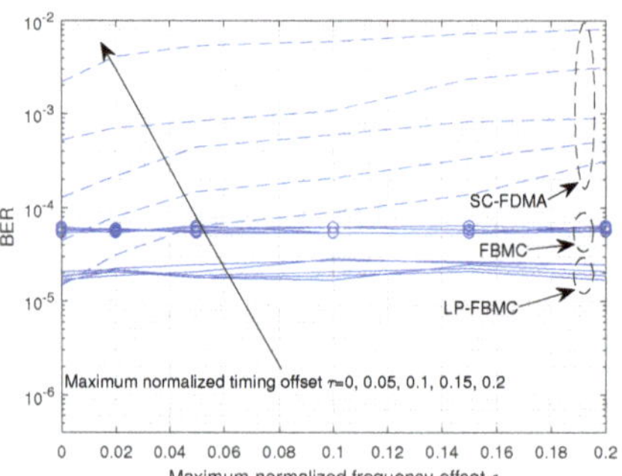

Figure 15. BER of the SC-FDMA, FBMC, and LP-FBMC according to timing/frequency offset, $\Delta f = 30$ kHz, $N = 64$, $F = 3.5$ GHz.

5. Conclusions

In this study, we intensively analysed the performance of SC-FDMA, FBMC, and LP-FBMC via field test using SDR devices in the real indoor wireless environments. As for indoor wireless channel, we tested the various sub-3.5 GHz bands when considering the different indoor wireless band allocations in the different countries. Basically, the benefits of LP-FBMC over FBMC and SC-FDMA, i.e., low PAPR, and the robustness against inter-user synchronization errors were confirmed in our field test. For instance, LP-FBMC achieves a BER of less than 1/10 of that of the FBMC in a practical indoor distance between the BS and user. In addition, no matter how large inter-user timing/frequency offset ranges are given in the considered region, LP-FBMC achieves the identical BER performance to that of zero offset case, whereas SC-FDMA severely degrades, even with slight offsets. Summing up the major observations of our field test, the BER performance gain of LP-FBMC over SC-FDMA is more increased for lower carrier frequency, larger subcarrier spacing, and larger timing/frequency offset among the users. In light of these results, we expect that LP-FBMC is a promising waveform solution for beyond 5G uplink waveforms in indoor sub-3.5 GHz Bands.

Author Contributions: Writing—review and editing, K.C.; writing—original draft preparation, D.N.; Implementation and field test, S.J.; project administration, W.-G.S.; resources, W.-G.S. All authors have read and agreed to the published version of the manuscript.

Funding: This research was supported by the MSIT (Ministry of Science and ICT), Korea, under the ITRC (I nformation Technology Research Center) support program (IITP-2020-2016-0-00313) supervised by the IITP (Institute for Information & communications Technology Planning & Evaluation. This research was supported by the National Research Foundation of Korea(NRF) grant funded by the Korea government (MSIT) (No. 2021R1A2C1010370).

Conflicts of Interest: The authors declare no conflict of interest.

References

1. Herminawan, F.; Higashino, T.; Okada, M. Investigation of Cross Modulation for SC-FDMA Signals in Radio over Fiber Mobile Link. In Proceedings of the IEEE Asia-Pacific Microwave Conference, Hong Kong, China, 8–11 December 2020.
2. Ihalainen, T.; Viholainen, A.; Stitz, T.H.; Renfors, M.; Bellanger, M. Filter bank based multi-mode multiple access scheme for wireless uplink. In Proceedings of the 17th European Signal Processing Conference, Glasgow, UK, 24–28 August 2009.
3. Schaich, F.; Wild, T. Waveform contenders for 5G—OFDM vs. FBMC vs. UFMC. In Proceedings of the 6th International Symposium on Communications, Control and Signal Processing (ISCCSP), Athens, Greece, 21–23 May 2014.
4. Kang, H.; Song, Y.; Kwon, D.; Kim, D. Key Techniques and Performance Comparison of 5G New Waveform Technologies. In Proceedings of the Korean Institute of Communications and Information Sciences (KICS), Seoul, Korea, 1 January 2016; Volume 41, pp. 142–155.
5. Zhang, N.; Wang, J.; Kang, G.; Liu, Y. Uplink Nonorthogonal Multiple Access in 5G Systems. *IEEE Commun. Lett.* **2016**, *20*, 458-461. [CrossRef]

6. Jang, S.; Na, D.; Choi, K. Comprehensive Performance Comparison between OFDM-based and FBMC-based Uplink Systems. In Proceedings of the International Conference on Information Networking (ICOIN), Barcelona, Spain, 7–10 January 2020.
7. Dore, J.; Cassiau, N.; Ktenas, D. Low complexity frequency domain carrier frequency offset compensation for uplink multiuser FBMC receiver. In Proceedings of the European Conference on Networks and Communications (EuCNC), Bologna, Italy, 23–26 June 2014.
8. Na, D.; Choi, K. Low PAPR FBMC. *IEEE Trans. Wirel. Commun.* **2018**, *17*, 182–193. [CrossRef]
9. Choi, K. Alamouti Coding for DFT Spreading-Based Low PAPR FBMC. *IEEE Trans. Wirel. Commun.* **2019**, *18*, 926–941. [CrossRef]
10. Na, D.; Choi, K. DFT Spreading-based Low PAPR FBMC with Embedded Side Information. *IEEE Trans. Commun.* **2019**, *68*, 1731–1745. [CrossRef]
11. Nandhakumar, P. A Review on Next Generation Mobile Communication-5G. *Sci. Bull. Electr. Eng. Fac.* **2019**, *19*, 16–20. [CrossRef]
12. Vitucci, E.M.; Yu, F.; Possenti, L.; Zoli, M.; Fuschini, F.; Barbiroli, M.; Degli-Esposti, V. A Study on Dual-Directional Mm-wave. In Proceedings of the European Conference on Antennas and Propagation, Krakow, Poland, 31 March–5 April 2019.
13. Bera, S.; Sarkar, S.K. Review on Indoor Channel Characterization for Future Generation Wireless Communications. *Adv. Commun. Devices Netw.* **2019**, *537*, 349–356.
14. Perez, J.R.; Torres, R.P.; Rubio, L.; Basterrechea, J.; Penarrocha, R.; Reig, J. Empirical characteriz- ation of the indoor radio channel for array antenna systems in the 3 to 4 GHz frequency band. *IEEE Access* **2019**, *7*, 94725–94736. [CrossRef]
15. Huang, F.; Tian, L.; Zheng, Y.; Zhang, J. Propagation characteristics of indoor radio channel from 3.5 GHz to 28 GHz. In Proceedings of the IEEE 84th Vehicular Technology Conference, Montreal, QC, Canada, 18–21 September 2016; pp. 1-5.
16. Nadal, J.; Nour, C.A.; Baghdadi, A.; Lin, H. Hardware prototyping of FBMC/OQAM baseband for 5G mobile communication. In Proceedings of the 2014 25th IEEE International Symposium on Rapid System Prototyping (RSP), New Delhi, India, 16–17 October 2014; pp. 72–77.
17. Barlee, K.W.; Stewart, R.W.; Crockett, L.H.; MacEwen, N.C. Rapid prototyping and validation of FS-FBMC dynamic spectrum radio with simulink and ZynqSDR. *IEEE Open J. Commun. Soc.* **2020**, *2*, 113–131. [CrossRef]
18. Lee, J.; Tejedor, E.; Ranta-aho, K.; Wang, H.; Lee, Ky.; Semaan, E.; Mohyeldin, E.; Song, J. Spectrum for 5G: Global Status, Challenges, and Enabling Technologies. *IEEE Commun. Mag.* **2018**, *56*, 12–18. [CrossRef]
19. Riback, M.; Medbo, J.; Berg, J.; Harrysson, F.; Asplund, H. Carrier Frequency Effects on Path Loss. In Proceedings of the IEEE 63rd Vehicular Technology Conference, Melbourne, Australia, 11–12 May 2006.
20. Farhang-Boroujeny, B. OFDM Versus Filter Bank Multicarrier. *IEEE Signal Process. Mag.* **2011**, *28*, 92–112. [CrossRef]

Article

Improved Pair-Wise Detections of Differential Quasi-Orthogonal Space-Time Modulation with Four Transmit Antennas

Hojun Kim [1], Yulong Shang [2], Seunghyeon Kim [3] and Taejin Jung [3,*]

1 R&D Team, Networks Business, Samsung Electronics, Suwon 16677, Korea; friendlyguy2319@gmail.com
2 School of Electrical and Information Engineering, Jiangsu University of Technology, Changzhou 213000, China; shangyulong@jsut.edu.cn
3 Department of ICT Convergence System Engineering, Chonnam National University, Gwangju 61186, Korea; 167709@live.jnu.ac.kr
* Correspondence: tjjung@chonnam.ac.kr; Tel.: +82-62-530-0722

Abstract: In this paper, we propose new complex and real pair-wise detection for conventional differential space–time modulations based on quasi-orthogonal design with four transmit antennas for general QAM. Since the new complex and real pair-wise detections allow the independent joint ML detection of two complex and real symbol pairs, respectively, the decoding complexity is the same as or lower than conventional differential detections. Simulation results show that the proposed detections exhibit almost identical performance with an optimum maximum-likelihood receiver, as well as improved performance compared with conventional pair-wise detections, especially for higher modulation order.

Keywords: space time codes; differential space time modulation; differential detection; pair-wise detection; maximum likelihood detection

Citation: Kim, H.; Shang, Y.; Kim, S.; Jung, T. Improved Pair-Wise Detections of Differential Quasi-Orthogonal Space-Time Modulation with Four Transmit Antennas. *Electronics* **2021**, *10*, 1675. https://doi.org/10.3390/electronics10141675

Academic Editor: Giovanni Andrea Casula

Received: 7 April 2021
Accepted: 8 July 2021
Published: 13 July 2021

1. Introduction

So-called quasi-orthogonal (QO) design, adopted in coherent space–time codes (STBCs) [1–3], enjoys some preferable features of full spatial diversity gain as well as simplified maximum-likelihood (ML) detection based on complex or real pair-wise symbols for any type of signal constellation. These QO-STBCs have been developed to be applicable not only to 4G long-term evolution (LTE) [4], but also to a 5G new radio (NR) communication system [5,6] that is currently commercializing beyond standardization.

However, for so-called differential space–time modulations (DSTMs) [7,8] based on the QO design, hereafter referred to as QO-DSTM, the efficient pair-wise ML detection applied in the conventional QO-STBCs is no longer available at a receiver when using general QAM. This is mainly since power-normalization with a constraint of fixed total transmit energy in a transmitter inevitably generates dependencies among all differentially modulated signals on each other. Hence, [7,8] presents alternative pair-wise, but not ML, detections showing degraded performance compared to the ML decoding. Furthermore, ref. [9] does not exhibit significant performance loss unlike [7,8], but has a critical disadvantage due to its decoding complexity greatly increasing as the modulation order increases.

For this reason, in this paper, new complex and real pair-wise detections for the conventional QO-DSTMs [7,8] with four transmit antennas were proposed for general QAM without additional operation. A key feature in the proposed detections is that when decoding a given complex or real symbols pair, all the other pairs' variant power values contained in an ML metric are simply replaced by or estimated to be their constant mean values. This mean-based estimation effectively cuts off the dependencies between the given differential symbols which pair with all the other pairs, thus enabling the independent joint detection of two complex or real symbols, such as the conventional ones [7,8], with the resulting performances much closer to ML decoding.

2. Conventional QO-DSTMs and Differential Pair-Wise Detections

The conventional QO-DSTMs [7,8] with four transmit antennas can be constructed by serially concatenating the coherent QO code with differential encoding, as shown in Figure 1.

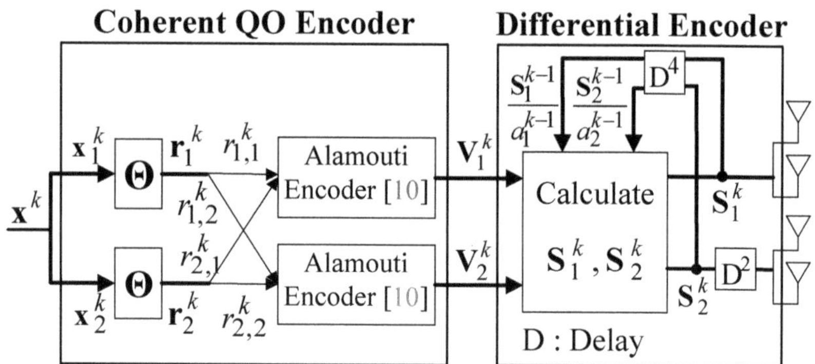

Figure 1. Block diagram of conventional QO-DSTM [7,8] with four transmit antennas.

The transmitter of Figure 1 first encodes a kth input vector $\mathbf{x}^k = [\mathbf{x}_1^k, \mathbf{x}_2^k]^T$ of length 4 through the conventional coherent QO encoder [2,3] consisting of precoder $\boldsymbol{\Theta}$ and Alamouti encoder [10], resulting in:

$$\mathbf{V}_1^k = \frac{1}{\sqrt{2}}\begin{bmatrix} r_{1,1}^k & r_{2,1}^k \\ -\left(r_{2,1}^k\right)^* & \left(r_{1,1}^k\right)^* \end{bmatrix}, \tag{1}$$

$$\mathbf{V}_2^k = \frac{1}{\sqrt{2}}\begin{bmatrix} r_{1,2}^k & r_{2,2}^k \\ -\left(r_{2,2}^k\right)^* & \left(r_{1,2}^k\right)^* \end{bmatrix} \tag{2}$$

where $\mathbf{r}_i^k = \left[r_{i,1}^k r_{i,2}^k\right]^T = \boldsymbol{\Theta}\mathbf{x}_i^k$ with an unitary precoder $\boldsymbol{\Theta}$. Notice that the precoder $\boldsymbol{\Theta}$ is chosen so that the precoded vector $\mathbf{r}_i^k$ has different values for any distinct $\mathbf{x}_i^k$ [2,3], and thus is mapped to $\mathbf{x}_i^k$ in a one-to-one relationship. Then, each $\mathbf{V}_l^k$ is differentially modulated with an iterative fashion as follows:

$$\mathbf{S}_l^0 = \mathbf{I}_2,\ \mathbf{S}_l^k = \frac{\mathbf{V}_l^k}{a_l^{k-1}}\mathbf{S}_l^{k-1}, k \geq 1 \tag{3}$$

where $a_l^{k-1} = \sqrt{\frac{1}{2}\left(|r_{1,l}^{k-1}|^2 + |r_{2,l}^{k-1}|^2\right)}$ is a power-normalization factor to satisfy $\mathbf{S}_l^{k-1}\left(\mathbf{S}_l^{k-1}\right)^H = \left(a_l^{k-1}\right)^2\mathbf{I}_2$ with an 2×2 identity matrix $\mathbf{I}_2$. $(\bullet)^H$ denotes a Hermitian operator. The differentially modulated $\mathbf{S}_l^k$ are finally transmitted through four transmit antennas in a time-multiplexed form as shown in Figure 1, and finally arrive at a receiver through 4×1 independent and identical MIMO fading channels.

Assuming a receive vector $\mathbf{y}_l^k = [y_{l,1}^k y_{l,2}^k]^T$ corresponding to $\mathbf{S}_l^k$, Zhu's QO-DSTM [7] presents a near-ML complex pair-wise detection of $\mathbf{x}_i^k$, where $i = 1,2$ for an optimal complex precoder $\mathbf{\Theta}_C = \frac{1}{\sqrt{2}}\begin{bmatrix} 1 & e^{j\pi/4} \\ 1 & -e^{j\pi/4} \end{bmatrix}$, given as

$$\hat{\mathbf{x}}_i^k = \min_{\mathbf{r}_i^k} \sum_{l=1}^{2} \left[\frac{||\mathbf{y}_l^{k-1}||^2}{2a_l^{k-1}} |r_{i,l}^k|^2 - \left\{ \left\langle \mathbf{y'}_{l,i}^{k-1}, \mathbf{y'}_{l,1}^k \right\rangle r_{i,l}^k \right\}^R \right] \quad (4)$$

where $\mathbf{y'}_{l,1}^{k-1} = [(y_{l,1}^{k-1})^* y_{l,2}^{k-1}]^T$ and $\mathbf{y'}_{l,2}^{k-1} = [(y_{l,2}^{k-1})^* - y_{l,1}^{k-1}]^T$. $(\bullet)^T$ is a transpose operator and $\langle a, b \rangle = a^H b$.

Furthermore, in order to further reduce the decoding complexity of (4), Chang's QO-DSTM [8] uses a real precoder $\mathbf{\Theta}_R = \begin{bmatrix} \cos\theta & \sin\theta \\ -\sin\theta & \cos\theta \end{bmatrix}$ with $\theta = \pi/4 + 13.28°$, producing real pair-wise detection of $\left(\mathbf{x}_i^k\right)^R$ and $\left(\mathbf{x}_i^k\right)^I$, $i = 1,2$, given as

$$\left(\hat{\mathbf{x}}_i^k\right)^R = \min_{(\mathbf{r}_i^k)^R} \sum_{l=1}^{2} \left[\frac{||\mathbf{y}_l^{k-1}||^2}{2a_l^{k-1}} \left|\left(r_i^k\right)^R\right|^2 \mp \left\{ \left\langle \mathbf{y'}_{i,l}^{k-1}, \mathbf{y'}_l^k \right\rangle^R \left(r_{i,l}^k\right)^R \right\} \right], \quad (5)$$

$$\left(\hat{\mathbf{x}}_i^k\right)^I = \min_{(\mathbf{r}_i^k)^I} \sum_{l=1}^{2} \left[\frac{||\mathbf{y}_l^{k-1}||^2}{2a_l^{k-1}} \left|\left(r_i^k\right)^I\right|^2 \mp \left\{ \left\langle \mathbf{y'}_{i,l}^{k-1}, \mathbf{y'}_l^k \right\rangle^I \left(r_{i,l}^k\right)^I \right\} \right]. \quad (6)$$

In (4)–(6), $(\bullet)^R$ and $(\bullet)^I$ denote real and imaginary parts, respectively.

3. New Pair-Wise Detections

Before deriving new pair-wise decoding, we first rearrange the exact ML decoding ([3], Equation (33)) in a vector form as follows:

$$\begin{aligned}
\hat{\mathbf{x}}^k &= \min_{\mathbf{x}^k} \sum_{l=1}^{2} \left(\begin{array}{c} \frac{\sigma^2 + \left(a_l^{k-1}\right)^2}{q_l} ||\mathbf{y}_l^k||^2 + \frac{\sigma^2 + \left(a_l^k\right)^2}{q_l} ||\mathbf{y}_l^{k-1}||^2 \\ -\frac{2a_l^{k-1}}{q_l} \left[\left(\mathbf{y}_l^{k-1}\right)^H \mathbf{V}_l^k \mathbf{y}_l^k \right]^R + 2\sigma^2 \ln(q_l) \end{array} \right) & (7) \\
&\approx \min_{\mathbf{x}^k} \left(\left\| \frac{a_1^{k-1}\mathbf{y}_1^k}{\mu_1} - \frac{\mathbf{V}_1^k \mathbf{y}_1^{k-1}}{\mu_1} \right\|^2 + \left\| \frac{a_2^{k-1}\mathbf{y}_2^k}{\mu_2} - \frac{\mathbf{V}_2^k \mathbf{y}_2^{k-1}}{\mu_2} \right\|^2 \right) & (8) \\
&= \min_{\mathbf{x}^k} \left\| \begin{bmatrix} a_1^{k-1} y_{1,1}^k / \mu_1 \\ a_1^{k-1} \left(y_{1,2}^k\right)^* / \mu_1 \\ a_2^{k-1} y_{2,1}^k / \mu_2 \\ a_2^{k-1} \left(y_{2,2}^k\right)^* / \mu_2 \end{bmatrix} - \begin{bmatrix} \frac{\mathbf{Y}_1^{k-1}}{\mu_1} & \mathbf{0}_2 \\ \mathbf{0}_2 & \frac{\mathbf{Y}_2^{k-1}}{\mu_2} \end{bmatrix} \begin{bmatrix} r_{1,1}^k \\ r_{2,1}^k \\ r_{1,2}^k \\ r_{2,2}^k \end{bmatrix} \right\|^2 & (9)
\end{aligned}$$

where $q_l = \sigma^2 + \left(a_l^{k-1}\right)^2 + \left(a_l^k\right)^2$ with $\sigma^2 = 1/\text{SNR}$, $\mu_l = \sqrt{\left(a_l^k\right)^2 + \left(a_l^{k-1}\right)^2}$ and $\mathbf{Y}_l^{k-1} = \begin{bmatrix} y_{l,1}^{k-1} & y_{l,2}^{k-1} \\ (y_{l,2}^{k-1})^* & -(y_{l,1}^{k-1})^* \end{bmatrix}$. The approximation of (8) uses an assumption of high SNR, i.e., $\sigma^2 \approx 0$ [7,8]. Furthermore, the equality of (9) comes from no change of magnitude of any conjugated signal.

Here, let us define a 4×4 unitary matrix $\mathbf{B} = \begin{bmatrix} \mathbf{Y}_1^{k-1}/\rho_1^{k-1} & \mathbf{0}_2 \\ \mathbf{0}_2 & \mathbf{Y}_2^{k-1}/\rho_2^{k-1} \end{bmatrix}$ with $\rho_l^{k-1} = \|\mathbf{y}_l^{k-1}\|$ to satisfy $\mathbf{B}^H\mathbf{B} = \mathbf{I}_4$. Then, by multiplying the left of (9) with $\mathbf{B}^H$, the ML metric can be written as the summation of two equations, including each $\mathbf{x}_i^k$ as follows:

$$\left\| \mathbf{B}^H \begin{bmatrix} a_1^{k-1} y_{1,1}^k / \mu_1 \\ a_1^{k-1} \left(y_{1,2}^k\right)^* / \mu_1 \\ a_2^{k-1} y_{2,1}^k / \mu_2 \\ a_2^{k-1} \left(y_{2,2}^k\right)^* / \mu_2 \end{bmatrix} - \mathbf{B}^H \begin{bmatrix} \frac{\mathbf{Y}_1^{k-1}}{\mu_1} & \mathbf{0}_2 \\ \mathbf{0}_2 & \frac{\mathbf{Y}_2^{k-1}}{\mu_2} \end{bmatrix} \begin{bmatrix} r_{1,1}^k \\ r_{2,1}^k \\ r_{1,2}^k \\ r_{2,2}^k \end{bmatrix} \right\|^2$$

$$= \left\| \begin{bmatrix} 1/\mu_1 & 0 & 0 & 0 \\ 0 & 1/\mu_1 & 0 & 0 \\ 0 & 0 & 1/\mu_2 & 0 \\ 0 & 0 & 0 & 1/\mu_2 \end{bmatrix} \left(\begin{bmatrix} z_{1,1}^k \\ z_{2,1}^k \\ z_{1,2}^k \\ z_{2,2}^k \end{bmatrix} - \begin{bmatrix} \rho_1^{k-1} & 0 & 0 & 0 \\ 0 & \rho_1^{k-1} & 0 & 0 \\ 0 & 0 & \rho_2^{k-1} & 0 \\ 0 & 0 & 0 & \rho_2^{k-1} \end{bmatrix} \begin{bmatrix} r_{1,1}^k \\ r_{2,1}^k \\ r_{1,2}^k \\ r_{2,2}^k \end{bmatrix} \right) \right\|^2 \quad (10)$$

$$= \left\| \begin{array}{l} \begin{bmatrix} 1/\mu_1 & 0 \\ 0 & 1/\mu_2 \end{bmatrix} \left(\begin{bmatrix} z_{1,1}^k \\ z_{1,2}^k \end{bmatrix} - \begin{bmatrix} \rho_1^{k-1} & 0 \\ 0 & \rho_2^{k-1} \end{bmatrix} \begin{bmatrix} r_{1,1}^k \\ r_{1,2}^k \end{bmatrix} \right) \\ + \begin{bmatrix} 1/\mu_1 & 0 \\ 0 & 1/\mu_2 \end{bmatrix} \left(\begin{bmatrix} z_{2,1}^k \\ z_{2,2}^k \end{bmatrix} - \begin{bmatrix} \rho_1^{k-1} & 0 \\ 0 & \rho_2^{k-1} \end{bmatrix} \begin{bmatrix} r_{2,1}^k \\ r_{2,2}^k \end{bmatrix} \right) \end{array} \right\|^2 \quad (11)$$

$$= \left\| \mathbf{\Lambda} \left(\mathbf{z}_1^k - \boldsymbol{\rho}^{k-1} \mathbf{\Theta} \mathbf{x}_1^k \right) \right\|^2 + \left\| \mathbf{\Lambda} \left(\mathbf{z}_2^k - \boldsymbol{\rho}^{k-1} \mathbf{\Theta} \mathbf{x}_2^k \right) \right\|^2 \quad (12)$$

where diagonal matrix $\mathbf{\Lambda} = \text{diag}(1/\mu_1, 1/\mu_2)$, $\mathbf{z}_i^k = [z_{i,1}^k z_{i,2}^k]^T$ with $z_{i,l}^k = \left\{ a_l^{k-1} \left\langle \mathbf{y}'^{k-1}_{l,i}, \mathbf{y}'^{k}_{l,1} \right\rangle \right\} / \rho_l^{k-1}$ and $\boldsymbol{\rho}^{k-1} = \text{diag}\left(\rho_1^{k-1}, \rho_2^{k-1}\right)$. The equality of (10) uses an energy conserving property of the unitary matrix $\mathbf{B}^H$.

From (12), we can see that if all μ_l in $\mathbf{\Lambda}$ are independent of $\mathbf{x}^k$, each $\mathbf{x}_i^k$ can be separately decoded by minimizing $\left\| \mathbf{\Lambda} \left(\mathbf{z}_i^k - \boldsymbol{\rho}^{k-1} \mathbf{\Theta} \mathbf{x}_i^k \right) \right\|^2$. Unfortunately, the values $\mu_l = \sqrt{\frac{1}{2}\left(|r_{1,l}^k|^2 + |r_{2,l}^k|^2 \right) + \left(a_l^{k-1}\right)^2}$ in $\mathbf{\Lambda}$ are variant with respect to $r_{i,l}^k$ or all the input QAM signals $\mathbf{x}^k$, implying the unfeasibility of pair-wise ML detection at the receiver. Specifically, in order to decode $\mathbf{x}_1^k$, we need to know the two exact values of $|r_{2,1}^k|^2$ and $|r_{2,2}^k|^2$ in μ_l containing the other signals $\mathbf{x}_2^k$. Conversely, for decoding $\mathbf{x}_2^k$, the two values $|r_{1,1}^k|^2$ and $|r_{1,2}^k|^2$ corresponding to the other $\mathbf{x}_1^k$ need to be known. Considering that μ_l are normalizing terms including a_l^k and a_l^{k-1}, we conclude that the dependency between two symbol pairs $\mathbf{x}_1^k$ and $\mathbf{x}_2^k$ indeed originates from the power-normalization with the constraint of total transmit power in (3), thus implying that this dependency will be not avoidable for the general QAM.

Hence, in order to make separate detections of $\mathbf{x}_i^k$ possible, it should be done to break up the dependency relationship between $\mathbf{x}_1^k$ and $\mathbf{x}_2^k$ in μ_l. For this goal, when decoding $\mathbf{x}_1^k$, we simply replace or estimate the values of the other $|r_{2,1}^k|^2$ by their mean values, i.e., $E\left\{|r_{2,1}^k|^2\right\} = E\left\{|r_{2,2}^k|^2\right\} = 1$, and also when decoding $\mathbf{x}_2^k$, $E\left\{|r_{1,1}^k|^2\right\} = E\left\{|r_{1,2}^k|^2\right\} = 1$. In this way, the decoding of $\mathbf{x}_1^k$ can be performed only by using $r_{1,l}^k$ elements containing $\mathbf{x}_1^k$ and the same can be done for decoding $\mathbf{x}_2^k$ only by using $r_{2,l}^k$ elements.

Hence, this simple mean-based estimation produces a new complex pair-wise decoding of $\mathbf{x}_i^k$, given as

$$\hat{\mathbf{x}}_i^k = \min_{\mathbf{x}_i^k} \left\| \hat{\mathbf{\Lambda}}_i \left(\mathbf{z}_i^k - \boldsymbol{\rho}^{k-1} \mathbf{\Theta} \mathbf{x}_i^k \right) \right\|^2, i = 1, 2 \quad (13)$$

where $\hat{\mathbf{\Lambda}}_i = \mathrm{diag}(1/\hat{\mu}_{i,1}, 1/\hat{\mu}_{i,2})$ with $\hat{\mu}_{i,l} = \sqrt{\frac{1}{2}\left(|r_{i,l}^k|^2 + 1\right) + (a_l^{k-1})^2}$.

Moreover, for a real $\mathbf{\Theta}$, the mean-based estimation can be applied to $\hat{\mu}_{i,l} = \sqrt{\frac{1}{2}\left(|(r_{i,l}^k)^R|^2 + |(r_{i,l}^k)^I|^2 + 1\right) + (a_l^{k-1})^2}$ in (13) one more time for the separate decoding of $\left(\mathbf{x}_i^k\right)^R$ and $\left(\mathbf{x}_i^k\right)^I$, i.e., $E\left\{|(r_{i,l}^k)^I|^2\right\} = \frac{1}{2}, \forall l$ when decoding $\left(\mathbf{x}_i^k\right)^R$ and also $E\left\{|(r_{i,l}^k)^R|^2\right\} = \frac{1}{2}, \forall l$ when decoding $\left(\mathbf{x}_i^k\right)^I$, resulting in:

$$\left(\hat{\mathbf{x}}_i^k\right)^R = \min_{\left(\mathbf{x}_i^k\right)^R} \left\| \hat{\mathbf{\Lambda}}_i^R \left[\left(\mathbf{z}_i^k\right)^R - \rho^{k-1} \mathbf{\Theta} \left(\mathbf{x}_i^k\right)^R \right] \right\|^2, i = 1, 2 \tag{14}$$

$$\left(\hat{\mathbf{x}}_i^k\right)^I = \min_{\left(\mathbf{x}_i^k\right)^I} \left\| \hat{\mathbf{\Lambda}}_i^I \left[\left(\mathbf{z}_i^k\right)^I - \rho^{k-1} \mathbf{\Theta} \left(\mathbf{x}_i^k\right)^I \right] \right\|^2, i = 1, 2 \tag{15}$$

where $\hat{\mathbf{\Lambda}}_i^R = \mathrm{diag}\left(1/\hat{\mu}_{i,1}^R, 1/\hat{\mu}_{i,2}^R\right)$ with $\hat{\mu}_{i,l}^R = \sqrt{\frac{1}{2}\left(|(r_{i,l}^k)^R|^2 + \frac{3}{2}\right) + (a_l^{k-1})^2}$ and $\hat{\mathbf{\Lambda}}_i^I = \mathrm{diag}\left(1/\hat{\mu}_{i,1}^I, 1/\hat{\mu}_{i,2}^I\right)$ with $\hat{\mu}_{i,l}^I = \sqrt{\frac{1}{2}\left(|(r_{i,l}^k)^I|^2 + \frac{3}{2}\right) + (a_l^{k-1})^2}$.

Obviously, the new pair-wise detections of (13)–(15) are not equal to the ML decoding of (12), but exhibit performances within only about 0.5 dB compared to the ML receiver for all simulation cases, which will be shown in the following simulation results.

Notice that with some manipulations, the conventional detection of (4) can be shown to be equal to the new one of (13) setting $\hat{\mu}_{i,l}$ to be $\sqrt{a_l^{k-1}}$, i.e., $\hat{\mathbf{\Lambda}}_i = \mathrm{diag}\left(\sqrt{a_1^{k-1}}, \sqrt{a_2^{k-1}}\right)$, given as

$$\begin{aligned}
&\min_{\mathbf{x}_i^k} \left\| \hat{\mathbf{\Lambda}}_i \left(\mathbf{z}_i^k - \rho^{k-1} \mathbf{\Theta} \mathbf{x}_i^k\right) \right\|^2 \\
&= \min_{\mathbf{r}_i^k} \left(\left\| \hat{\mathbf{\Lambda}}_i \rho^{k-1} \mathbf{r}_i^k \right\|^2 - 2\left\{ \left\langle \hat{\mathbf{\Lambda}}_i \mathbf{z}_i^k, \hat{\mathbf{\Lambda}}_i \rho^{k-1} \mathbf{r}_i^k \right\rangle \right\}^R \right) &(16) \\
&= \min_{\mathbf{r}_i^k} \sum_{l=1}^{2} \left[\frac{(\rho_1^{k-1})^2}{a_l^{k-1}} |r_{i,l}^k|^2 - 2\left\{ \left\langle \mathbf{y}'^{k-1}_{l,i}, \mathbf{y}'^{k}_{l,1} \right\rangle r_{i,l}^k \right\}^R \right] &(17)
\end{aligned}$$

where the equality of (16) is from that $\hat{\mathbf{\Lambda}}_i$ is absolutely irrelevant to the current input signals $\mathbf{x}_i^k$ or $\mathbf{r}_i^k$. Following the same derivations, (5) and (6) can also be proven to be equal to the new ones of (14) and (15), respectively, setting $\hat{\mu}_{i,l}^{R(I)}$ to be $\sqrt{a_l^{k-1}}$. This means that the conventional detections can be seen as the ones to cut off the dependency between two $\mathbf{x}_i^k$ or four $\left(\mathbf{x}_i^k\right)^{R(I)}$ from each other by estimating μ_l to be simply $\sqrt{a_l^{k-1}}$. Moreover, since the proposed methods are the same symbol by symbol decoding as the conventional methods except for the m_l computation, performance improvement can be expected without increasing the same decoding complexity.

Defining $\varepsilon(\hat{\mu}_l) = E\left\{(\mu_l - \hat{\mu}_l)^2\right\}$, Table 1 compares the accuracies of $\hat{\mu}_l$s used in the conventional and new detections for the QO-DSTMs [7,8] with perfect knowledge of a_l^{k-1}.

From Table 1, the estimation errors of $\hat{\mu}_l$ in the new detections are shown to be greatly lower than those of the convention detections. Furthermore, the gap between both detections becomes slightly larger as increasing the modulation order. This is obvious since the conventional detections do not consider the values of current QAM input signals $\mathbf{x}^k$ for estimating μ_l unlike new ones, and thus the estimation error increases much more for a higher modulation order.

Table 1. Comparisons of $\varepsilon(\hat{\mu}_l)$ for the conventional and new detections.

Mod.	Θ_C [7]		Θ_R [8]	
	$\varepsilon\left(\sqrt{a_l^{k-1}}\right)$, (4)	$\varepsilon(\hat{\mu}_l^i)$, (13)	$\varepsilon\left(\sqrt{a_l^{k-1}}\right)$, (5), (6)	$\varepsilon\left(\hat{\mu}_{l,R(l)}^i\right)$, (14), (15)
16-QAM	7.462	0.195	2.679	0.359
64-QAM	44.593	0.866	6.558	0.754
256-QAM	263.204	2.898	16.042	1.381

4. Simulation Results

All simulations are done based on an independent data frame consisting of 128 blocks. Each block has four QAM symbols. For propagation channel models, it is assumed that the four MIMO channel gains have independent and identical Rayleigh distributions, and also are constant during each frame with independent distribution. Furthermore, in all decodings, we use a_1^{k-1} and a_2^{k-1} calculated from previously detected symbols.

Figure 2 shows average bit error rates (BERs) of the new and conventional complex pair-wise detections for the Zhu's QO-DSTM [7] using Θ_C. For the comparison of performances, the ML results of (12) are also included. Firstly, the new pair-wise detection is shown to achieve more improved performance compared to the conventional one for all simulation cases. The performance gain is much larger especially as the modulation order increases. Specifically, for 16, 64 and 256 QAMs, the respective SNR gains at BER = 10^{-4} are about 0.8, 1.0, 1.2 dBs. This is mainly because the new detection is performed based on more accurately estimated $\hat{\mu}_l$ compared to the conventional one, especially for a higher modulation order, as shown in Table 1. Furthermore, we note that the proposed detection shows an SNR loss of less than only 0.2 dB compared to the ML decoding for all cases.

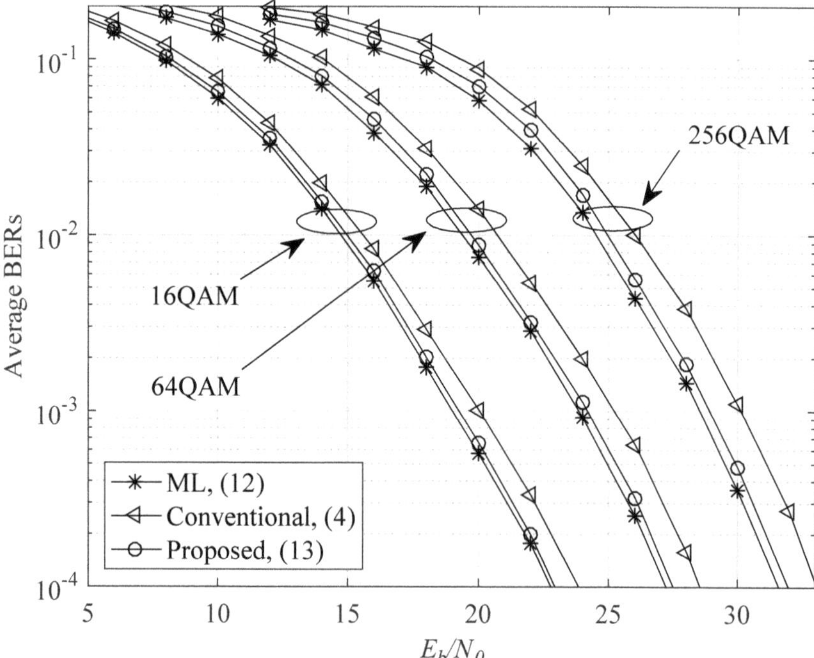

Figure 2. Average BERs for the QO-DSTM [7] with Θ_C.

Figure 3 shows the average BERs of the new and conventional real pair-wise detections for the Chang's QO-DSTM [8] using Θ_R. Here, the ML results are also included for comparing performances. First the performance trends in Figure 3 are almost the same as those of Figure 2 with the identical reasons as in Table 1. Specifically, the respective SNR gains for 16, 64 and 256 QAMs at BER = 10^{-4} are about 0.3, 0.6, 0.9 dBs. Furthermore, the proposed detection shows an SNR loss of less than 0.5 dB compared to the ML decoding for all cases.

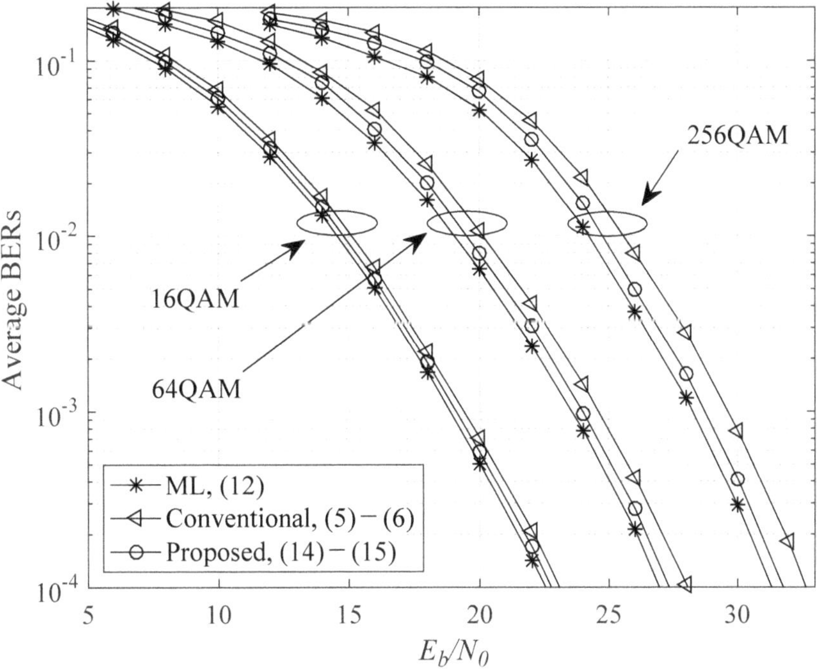

Figure 3. Average BERs for the QO-DSTM [8] with Θ_R.

Comparing the simulation results in Figures 2 and 3, we can see that the performance gain comparing to the conventional system is more significant when using the complex precoder in Figure 2. This is mainly due to the difference in estimation errors between the proposed and conventional methods listed in Table 1. Namely, when using the complex precoder, the gap of estimation errors for μ_l between these two methods are larger than those when using the real precoder for all modulation cases, as shown in Table 1.

5. Conclusions

In this paper, we proposed new complex and real pair-wise detections for the conventional QO-DSTMs with four transmit antennas for general QAM. The proposed detections exhibit a greatly improved performance compared to the conventional ones, especially, for a higher modulation order and also a performance almost identical with the ML decoding.

Hence, considering decoding the complexity and error performances, the new pair-wise detections are much more attractive for demodulating the QO-DSTMs. The mean-based estimation used in the proposed detections can indeed be applied to any other DSTMs based on amicable orthogonal [11] or QO [12] space-time codes with more than four transmit antennas and general QAM. In addition, it can be applicable for the other differential modulation systems in such as radio frequency technology [13], underwater communications [14], heterogeneous networks [15] and wireless sensor networks [16], which can also be applied to the artificial intelligence field [17] which is in the spotlight these days.

Author Contributions: Conceptualization, H.K. and T.J.; methodology, H.K.; software, S.K.; validation, Y.S. and S.K.; formal analysis, H.K. and T.J.; investigation, Y.S.; writing—original draft preparation, H.K.; writing—review and editing, H.K., Y.S. and T.J.; visualization, Y.S.; supervision, T.J.; project administration, T.J.; funding acquisition, T.J. All authors have read and agreed to the published version of the manuscript.

Funding: This research was supported by the BK21 FOUR Program(Fostering Outstanding Universities for Research, 5199991714138) funded by the Ministry of Education(MOE, Korea) and National Research Foundation of Korea(NRF).

Conflicts of Interest: The authors declare no conflict of interest.

References

1. Jafarkhani, H. A quasi-orthogonal space-time block code. *IEEE Trans. Commun.* **2001**, *49*, 1–4. [CrossRef]
2. Su, W.; Xia, G. Signal constellations for quasi-orthogonal space-time block codes with full diversity. *IEEE Trans. Inform. Theory* **2004**, *50*, 2331–2347. [CrossRef]
3. Chae, C.; Jung, T.; Hwang, I. Design of new quasi-orthogonal STBC with minimum decoding complexity for four transmit antennas. *IEICE Trans. Commun.* **2008**, *91-B*, 3368–3370. [CrossRef]
4. Alam, S.; GoangSeog, C.; Shajeel, I. Performance analysis of orthogonal space-time codes applicable to 4G LTE communications. In Proceedings of the 2015 International Symposium on Consumer Electronics (ISCE), Madrid, Spain, 24–26 June 2015.
5. Gurpreet, K.; Navjot, K.; Lavish, K. Performance Comparison of large MIMO Systems Using Quasi Orthogonal Space Time Block Coding Through AWGN and Rayleigh Channels by Zero Forcing Receivers. In Proceedings of the 2018 International Conference on Intelligent Circuits and Systems (ICICS), Phagwara, India, 19–20 April 2018.
6. Liu, C.; Xia, X.G.; Li, Y.; Gao, X.; Zhang, H. Omnidirectional Quasi-Orthogonal Space–Time Block Coded Massive MIMO Systems. *IEEE Commun. Let.* **2019**, *23*, 1621–1625. [CrossRef]
7. Zhu, Y.; Jafarkhani, H. Differential modulation based on quasi-orthgonal codes. *IEEE Trans. Wirel. Commun.* **2005**, *4*, 3018–3030.
8. Chang, T.; Hua, Y.; Sadler, B. A New Design of Differential Space-Time Block code Allowing Symbol-Wise Decoding. *IEEE Trans. Wirel. Commun.* **2007**, *6*, 3197–3201. [CrossRef]
9. Morsi, R.; Linduska, A.; Lindner, J. Full-diversity minimum decoding complexity differential quasi-orthogonal STBC. In Proceedings of the 2011 IEEE Global Telecommunications Conference (GLOBECOM), Houston, TX, USA, 5–9 December 2011.
10. Alamouti, A. A simple transmit diversity technique for wireless communications. *IEEE J. Sel. Areas Commun.* **1998**, *16*, 1451–1458. [CrossRef]
11. Chen, Z.; Zhu, G.; Qu, D.; Liu, M. General Differential Space-Time Modulation. In Proceedings of the 2003 IEEE Global Telecommunications Conference (GLOBECOM), San Francisco, CA, USA, 1–5 December 2003; pp. 282–286.
12. Song, L.; Burr, A. General differential modulation scheme for quasi-orthogonal space–time block codes with partial or full transmit diversity. *IET Commun.* **2007**, *1*, 256–266. [CrossRef]
13. Chao, X.; Peichang, Z.; Rakshith, R.; Naoki, I.; Shinya, S.; Li, W.; Lajos, H. Finite-Cardinality Single-RF Differential Space-Time Modulation for Improving the Diversity-Throughput Tradeoff. *IEEE Trans. Commun.* **2019**, *67*, 318–335.
14. Fengzhong, Q.; Zhenduo, W.; Liuqing, Y. Differential Orthogonal Space-Time Block Coding Modulation for Time-Variant Underwater Acoustic Channels. *IEEE J. Ocean. Eng.* **2017**, *42*, 188–198.
15. Syed, S.M.; Daniel, C.; Ahmed, A.; Elvino, S.S.; Adao, S.; Atilio, G. Joint Space-Frequency Block Codes and Signal Alignment for Heterogeneous Networks. *IEEE Access* **2018**, *6*, 71099–71109.
16. Kanthimathi, M.; Amutha, R.; Bhavatharak, N. Performance Analysis of Multiple-Symbol Differential Detection based Space-Time Block Codes in Wireless Sensor Networks. In Proceedings of the 2019 International Conference on Vision Towards Emerging Trends in Communication and Networking (ViTECoN), Vellore, India, 30–31 March 2019.
17. Mehrtash, M.; Mostafa, M.; Masoud, A.; Yindi, J. Decision Directed Channel Estimation Based on Deep Neural Network k -Step Predictor for MIMO Communications in 5G. *IEEE J. Sel. Areas Commun.* **2019**, *37*, 2443–2456.

Article

Waveform Design for Space–Time Coded MIMO Systems with High Secrecy Protection †

Pingping Shang, Hyein Lee and Sooyoung Kim *

IT Convergence Research Center, Division of Electronics Engineering, Jeonbuk National University, 567 Baekje-daero, Deokjin-gu, Jeonju 54896, Jeollabuk-do, Korea; pingajiayou@jbnu.ac.kr (P.S.); leehyein96@jbnu.ac.kr (H.L.)

* Correspondence: sookim@jbnu.ac.kr

† This paper is an extended version of our paper published in ICTC 2020.

Received: 14 October 2020; Accepted: 17 November 2020; Published: 25 November 2020

Abstract: In this paper, we present a new secrecy-enhancing scheme for multi-input-multi-output (MIMO) systems using a space–time coding scheme. We adopt a quasi-orthogonal space–time block coding (QO-STBC) scheme that was originally designed to improve the performance of the MIMO system, and propose an efficient waveform design that can enhance the secrecy, as well as improve the error rate performance. Channel- and signal-dependent artificial interference (AI) is added to the proposed waveform, so that only a legitimate receiver can successfully retrieve information. We investigate the secrecy capacity of the proposed scheme, and demonstrate that the proposed scheme provides highly enhanced secrecy performance, compared to the conventional schemes. The performance simulation results reveal that the transmitted information can be properly extracted only at the legitimate receiver.

Keywords: multi-input-multi-output (MIMO); space time block coding; physical layer security (PLS); secrecy capacity

---

1. Introduction

Most technology development in wireless systems has mainly focused on the enhancement of spectral efficiency and/or power efficiency (error rate performance). The explosion of multiple-input multiple-output (MIMO) technology can be one of the representative examples that have largely contributed to enhancing spectral or power efficiency. Due to their inherent broadcasting nature, wireless communication systems are vulnerable to security and privacy protection. However, waveform design for the past generation systems has not focused mainly on the secrecy performance aspect. Although the classical cryptography protocols in the network layer may guarantee secure communications, the involved secret-key distribution and management processes are generally unaffordable, and are fragile to attacks, especially in wireless systems [1]. For this reason, security protection needs to be considered during the waveform design process along with power and spectral efficiency.

Physical layer security (PLS) schemes can be considered as one of the effective waveform design techniques to provide security protection, and a number of study results have been reported by exploiting the characteristics of wireless channels [2–4]. Because MIMO techniques are now almost mandatory in most wireless systems, PLS schemes combined with MIMO techniques have recently gained a lot of attention. As a measure of secrecy protection, Reference [5] derived the secrecy capacity of a MIMO system. Later, Reference [6] proved that perfect secrecy protection could theoretically be achieved for MIMO systems, and these results accelerated the development of PLS techniques for MIMO systems. Reference [7] considered utilization of spatial modulation (SM) and adding artificial noise (AN) as effective means to increase the secrecy capacity for MIMO systems. The essence of AN technology is to add a noise that is orthogonal to the legitimate channel, so that it can be null to the legitimate receiver, while presenting interference to the illegal receiver.

On the other hand, there have been efforts to utilize space–time block coding (STBC) schemes for PLS [8–13]. The Alamouti scheme was utilized for PLS [8–12]. As a cooperative multi-user transmission technique, Reference [8] proposed the idea of transmit antenna selection (TAS)-Alamouti scheme for PLS enhancement. Reference [9] proposed another cooperative Alamouti scheme, where Alamouti users collaborate with each other with an AN-aided technique to secure the transmissions. They demonstrated that the secrecy sum-capacity was improved by adding AN aligned with the null space of the legitimate channel matrix. The Alamouti scheme was also utilized in SM with AN, where SM was used to increase the secrecy performance while the Alamouti scheme was used to enhance the performance [10]. Reference [11] proposed a PLS scheme employing a coordinate interleaved orthogonal design, while Reference [12] proposed a transmission technique for PLS by applying the Alamouti coded non-orthogonal multiple access scheme.

Furthermore, secrecy outage probability was derived for a quasi-orthogonal STBC (QO-STBC) scheme that linearly combines two Alamouti codes [13]. The results in this research demonstrate the secrecy outage probability according to the power ratio of legitimate and illegal receivers. This QO-STBC scheme utilized a power scaling (PS) method, but the PS itself did not contribute to enhancing the secrecy performance. Instead, the PS was used to achieve the full rate and full diversity effect. Reference [14] also proposed a similar idea, presenting a new encoding matrix for QO-STBC to achieve full orthogonality of the channel matrix.

In this paper, we propose a new waveform design for the MIMO system using STBC, which is targeted at enhancing the secrecy, as well as the error rate performance. We modify the idea of combing AN with STBC, and adopt a QO-STBC scheme with linear detection capability [14]. The advantage of the QO-STBC scheme in [14] lies in the full orthogonality of the channel matrix, and thus it was termed a linear decoding QO-STBC (LD-QO-STBC) scheme. However, its disadvantage lies in the non-uniform signal power distribution. In the proposed scheme, we utilize this property, and add the artificial interference (AI) by considering non-uniform signal distribution. The waveform with time-varying AI is received at the legitimate receiver without any interference, while the added interference at the illegal receiver prevents proper detection.

We note that STBC schemes are widely used in modern wireless systems to improve performance. Even though the STBC schemes were investigated for PLS in the previous studies, they can achieve reasonable secrecy protection only with a limited condition, i.e., when the power strength of the legitimate receiver is sufficiently higher than that of the illegal receiver. An approach with AN can be used, but this requires additional power allocation. Therefore, the proposed PLS scheme can be efficiently utilized for many wireless systems equipped with STBC, without any performance degradation.

Compared to the conventional PLS schemes with STBC, we summarize the superiority of our work as follows. First, the proposed scheme can be utilized in the STBC MIMO systems with more than two transmit antennae, and thus it can achieve more diversity gain compared to the previous works with the Alamouti scheme [8–12]. Second, the proposed scheme does not allocate additional power to AN for security protection, and thus it is power efficient. The probability distribution of the LD-QO-STBC signal waveform is newly derived. By using the investigation results of the signal distribution, the power reallocation is dynamically performed across the multi transmit antennae. This power reallocation contributes to equalizing power distribution across the transmit antennae as well as transmit time. In addition, it imposes severe interference on the eavesdropper. The novelty of the proposed idea lies in using the new waveform for STBC MIMO systems with power reallocation. By virtue of the above two strong advantages, the proposed scheme can achieve perfect security protection even though the eavesdropper has almost the same power as the legitimate receiver.

The rest of this paper is organized as follows. Section 2 describes the related works with the proposed scheme. We first introduce the PLS scheme that combines the Alamouti-code with AN, and present the LD-QO-STBC scheme [14]. Section 3 presents the proposed waveform design for PLS using the LD-QO-STBC scheme. We first investigate the characteristics of the signal constellation of the LD-QO-STBC scheme. Afterwards, we propose a method to generate signal- and channel-dependent AI that can be perfectly canceled only at the legitimate receiver. Section 4 evaluates the secrecy capacity of the proposed scheme, and compares it with that of the conventional scheme. In addition, we present a bit error rate (BER) performance simulation results to demonstrate proper security protection. Finally, Section 5 draws the conclusions.

Notation: Bold lower case letters represent vectors, while bold upper case letters denote matrices. $\mathbf{A}^T$, $\mathbf{A}^H$, and $\|\mathbf{A}\|$ denote the transpose, Hermitian transpose, and Euclidean norm of matrix $\mathbf{A}$, respectively. The superscript $(\cdot)^*$ denotes the complex conjugate, while $*$ denotes the convolution operator.

2. Related Works

2.1. Alamouti Coded PLS with AN

The generic model for the PLS scheme, comprises a cooperative wireless network that consists of three nodes. Suppose that a legitimate transmitter node is referred to as Alice, which is the source node. The corresponding destination node is referred to as Bob, which is the legitimate receiver node. On the other hand, the third node, named Eve, is the passive eavesdropper node.

Figure 1 shows a block diagram for the Alamouti-coded PLS with AN [10]. In this system, it is assumed that Alice is equipped with three antennae, while Bob and Eve are equipped with only one antenna, respectively. At the transmitter, the bit stream emitted by a binary source is divided into a block containing one bit to select one antenna among Ant 1 and Ant 2 for SM, and the other blocks containing information bits to transmit. The selected antenna will form the first virtual antenna for the Alamouti scheme, while Ant 3 in Figure 1 will form the second virtual antenna for the 2×1 Alamouti scheme. Therefore, the first virtual antenna is dynamically changed by time. During the first symbol period, s_1 and s_2 which are mapped by using m bits for each will be transmitted through two antennae, respectively. In addition, AN will be injected for secrecy purpose. During the second symbol period, the Alamouti encoded symbols, $-s_2^*$ and s_1^*, are transmitted along with ANs through two antennae, respectively.

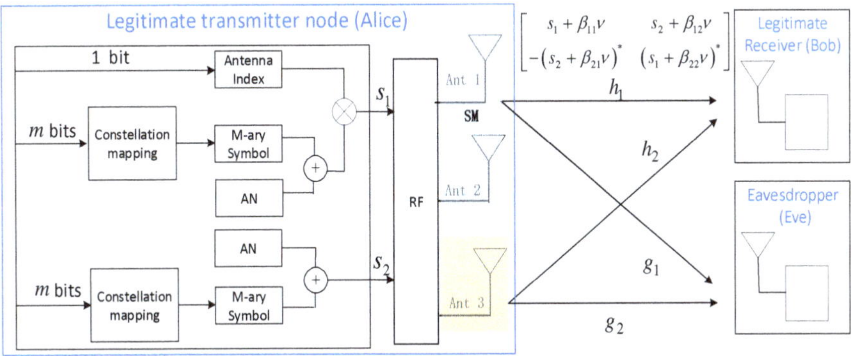

Figure 1. System model of the Alamouti-coded physical layer security (PLS) with artificial noise (AN).

It is assumed that the channel coefficients are constant within each transmission block, i.e., two symbol periods, and they are changed independently from block to block, as in the conventional

Alamouti scheme. The perfect channel state information (CSI) is assumed to be available at Bob and Eve, but it may not be available at Alice. The encoding matrix for the Alamouti code with AN, $\mathbf{X}_{\text{AN}}$ can be expressed as:

$$\mathbf{X}_{\text{AN}} = \begin{bmatrix} s_1 + \beta_{11} v & s_2 + \beta_{12} v \\ -(s_2 + \beta_{21} v)^* & (s_1 + \beta_{22} v)^* \end{bmatrix}, \quad (1)$$

where s_1 and s_2 are modulation symbols used for the Alamouti code during two symbol periods, and they are drawn from the M-ary quadrature amplitude modulation (QAM) constellation, where $M = 2^m$. In addition, v is the complex Gaussian AN with zero-mean and unit variance, and β_{ij}s are the coefficients of the v for the i-th timeslot at the j-th antenna. The coefficients β_{ij}s are designed to be nulled at the legitimate receiver. Therefore they should be computed by exploiting the CSI of the legitimate channel, which results in the following two possible sets of solutions for each symbol period, i.e., for the first time period:

$$\beta_{11} = h_2,\ \beta_{12} = -h_1,\ \text{or}\ \beta_{11} = -h_2,\ \beta_{12} = h_1, \quad (2)$$

and for the second time period:

$$\beta_{21} = h_2^*,\ \beta_{22} = h_1^*,\ \text{or}\ \beta_{21} = -h_2^*,\ \beta_{22} = -h_1^*, \quad (3)$$

where, h_1 and h_2 are the channel gains from the first and second virtual antennae at Alice to Bob, respectively.

With the above solution, the received signal at Bob during two symbol periods, $y_{\text{b}1}$ and $y_{\text{b}2}$ can be represented by:

$$\begin{aligned} y_{\text{b}1} &= h_1 s_1 + h_2 s_2 + n_{\text{b}1}, \\ y_{\text{b}2} &= -h_1 s_2^* + h_2 s_1^* + n_{\text{b}2}, \end{aligned} \quad (4)$$

where $n_{\text{b}1}$ and $n_{\text{b}2}$ are additive white Gaussian noise (AWGN) at Bob for the first and second symbol periods, respectively. Because the added ANs are perfectly canceled due to (2) and (3) at Bob, the detection can be successfully made as in the conventional 2×1 Alamouti scheme. On the other hand, the received signal at Eve will contain serious interference terms as follows:

$$\begin{aligned} y_{\text{e}1} &= g_1 (s_1 + \beta_{11} v) + g_2 (s_2 + \beta_{12} v) + n_{\text{e}1}, \\ y_{\text{e}2} &= -g_1 (s_2 + \beta_{21} v)^* + g_2 (s_1 + \beta_{22} v)^* + n_{\text{e}2}, \end{aligned} \quad (5)$$

where g_1 and g_2 are the channel gains from the first and second virtual antennae at Alice to Eve, respectively. In addition, $n_{\text{e}1}$ and $n_{\text{e}2}$ are AWGN at Eve for the first and second symbol periods, respectively.

2.2. Linear Decoding QO-STBC Scheme

After Alamouti first proposed the orthogonal STBC scheme with two transmit antennae [15], efforts were made to invent full-rate STBC schemes with a larger number of transmitting antennae. In an effort to increase the antenna size and achieve the full data rate, by loosening the orthogonality condition, several QO-STBC schemes have been proposed [16,17]. These QO-STBC schemes could achieve the full rate, but at the cost of higher decoding complexity, due to the non-perfect orthogonality of the encoding matrix.

In order to solve this problem, a new QO-STBC scheme providing a simple linear detection capability was proposed [14]. By applying the Givens rotation to the detection matrix of a conventional

QO-STBC scheme, new encoding matrices for three and four antenna systems were derived. The encoding matrix, $\mathbf{X}$, for four transmit antennae can be expressed as follows:

$$\mathbf{X} = \begin{bmatrix} s_1 + s_3 & s_2 + s_4 & s_3 - s_1 & s_4 - s_2 \\ -s_2^* - s_4^* & s_1^* + s_3^* & s_2^* - s_4^* & s_3^* - s_1^* \\ s_3 - s_1 & s_4 - s_2 & s_1 + s_3 & s_2 + s_4 \\ s_2^* - s_4^* & s_3^* - s_1^* & -s_2^* - s_4^* & s_1^* + s_3^* \end{bmatrix}. \tag{6}$$

As shown above, a signal transmitted for each time slot at each antenna has now become a linear combination of two modulation symbols. By assuming a single receive antenna, the received signal during four time slots can be expressed as

$$[y_1\ y_2\ y_3\ y_4]^T = \mathbf{X}\,[h_1\ h_2\ h_3\ h_4]^T + \mathbf{n}^T, \tag{7}$$

where h_i is the channel gain from the i-th transmit antenna to the receive antenna, and $\mathbf{n} = [n_1\ n_2\ n_3\ n_4]$ is an AWGN vector for the four time slots. For detection, the received signal vector can be represented by using the equivalent channel matrix, $\mathbf{H}$, as follows:

$$[y_1\ y_2^*\ y_3\ y_4^*]^T = \mathbf{H}\,[s_1\ s_2\ s_3\ s_4]^T + [n_1\ n_2^*\ n_3\ n_4^*]^T, \tag{8}$$

where

$$\mathbf{H} = \begin{bmatrix} h_1 - h_3 & h_2 - h_4 & h_1 + h_3 & h_2 + h_4 \\ h_2^* - h_4^* & h_3^* - h_1^* & h_2^* + h_4^* & -h_1^* - h_3^* \\ h_3 - h_1 & h_4 - h_2 & h_1 + h_3 & h_2 + h_4 \\ h_4^* - h_2^* & h_1^* - h_3^* & h_2^* + h_4^* & -h_1^* - h_3^* \end{bmatrix}. \tag{9}$$

The channel matrix $\mathbf{H}$ is orthogonal, although the encoding matrix $\mathbf{X}$ is quasi-orthogonal. For this reason, maximum likelihood (ML) decoding can be achieved via simple linear detection, as given by:

$$\hat{\mathbf{s}} = \mathbf{H}^H\,[y_1\ y_2^*\ y_3\ y_4^*]^T. \tag{10}$$

3. Secrecy-Enhancing LD-QO-STBC Scheme

3.1. Waveform Analysis for LD-QO-STBC

In order to design an efficient waveform for secrecy enhancement, we first investigate the signal constellation of the LD-QO-STBC scheme. By letting combinations of modulation symbols be used in (6) as follows:

$$\begin{aligned} x_1 &= s_1 + s_3, \quad x_2 = s_2 + s_4, \\ x_3 &= s_3 - s_1, \quad x_4 = s_4 - s_2, \end{aligned} \tag{11}$$

the encoding matrix $\mathbf{X}$ in (6) can be represented as:

$$\mathbf{X} = \begin{bmatrix} x_1 & x_2 & x_3 & x_4 \\ -x_2^* & x_1^* & -x_4^* & x_3^* \\ x_3 & x_4 & x_1 & x_2 \\ -x_4^* & x_3^* & -x_2^* & x_1^* \end{bmatrix}. \tag{12}$$

Because each element of $\mathbf{X}$ is a linear combination of any possible values of s_i and s_j, the waveforms of s_i and x_i will be different. In order to investigate the signal distributions, we denote constellation symbol sets used for s_i and x_i as $a \in O_a$ and $b \in O_b$, respectively. We note that for any Gray-coded QAM constellation symbol mapping, s_i, $-s_i$, s_i^*, and $-s_i^*$ are all in the same set O_a, due to its symmetric constellation. Likewise, elements of $\mathbf{X}$ that are either x_i, $-x_i$, x_i^*, or $-x_i^*$, are in the same set O_b. Nevertheless, it is clear that O_b forms a different set from O_a.

Figure 2 demonstrates an example of waveform transformation from O_a to O_b when 4-QAM is used for a. In the figure, we denote a_i and b_i as one of the complex constellation symbols included in O_a and O_b, respectively. In addition, we derive the probability density function (PDF) of b, and show that it is not a uniform distribution. For any M-ary QAM, we can assume that without loss of generality, the PDF of a follows a uniform distribution. That is,

$$f_a(a) = 1/M, \ \ a \in O_a. \tag{13}$$

By assuming the independency of the real and imaginary parts of a,

$$f_{a^R}(a^R) = 1/\sqrt{M}, \ \ f_{a^I}(a^I) = 1/\sqrt{M}, \ \ a \in O_a, \tag{14}$$

where a^R and a^I are the real and imaginary parts of a complex value a, respectively.

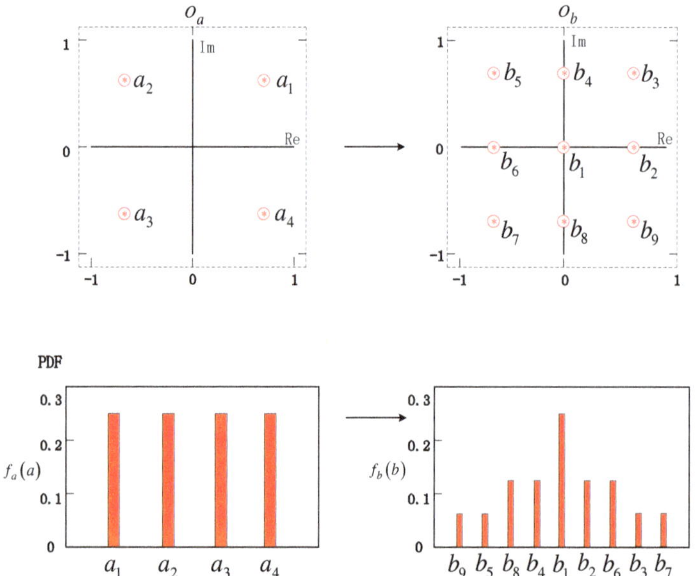

Figure 2. Signal constellations and PDFs for $a \in O_a$ and $b \in O_b$ used for s_i and x_i, respectively.

For any symbol b, we note that it is the sum of two independent random variables, so its PDF can be found by first taking the convolution of $f_{b^R}(b^R)$ and $f_{b^I}(b^I)$ in their real and imaginary parts, respectively. Therefore,

$$\begin{aligned} f_{b^R}(b^R) &= f_{a^R}(a^R) * f_{a^R}(a^R), \\ f_{b^I}(b^I) &= f_{a^I}(a^I) * f_{a^I}(a^I), \ \ a \in O_a, \end{aligned} \tag{15}$$

and finally using the independency of real and imaginary parts,

$$f_b(b) = f_{b^R}(b^R) f_{b^I}(b^I), \ \ b \in O_b. \tag{16}$$

Figure 2 compares the PDFs, $f_a(a)$ and $f_b(b)$ when 4-QAM is used for O_a. We find that the number of elements in O_b is greater than that in O_a, and that there is a null value $0 + j0$ in O_b, which is denoted by b_1 in the example of 4-QAM of Figure 2. Whatever M-ary QAM is used for O_a, this null value exists. We also find that the probability of having the null value of b is the maximum among the other possible values, i.e., $f_b(b_1)$ is the maximum amongst, and Figure 2 also demonstrates this. These findings will be applicable to higher-order M-ary QAM. The resultant O_b from the LD-QO-STBC scheme will incur zero-crossing amongst non-zero modulation symbol values.

3.2. The Design of Artificial Interference

From the investigation in the previous section, we note that the probability of having the null value of x_i, i.e., $x_i = 0 + j0$ is the maximum amongst all possible values in O_b. We also note from (6) that whenever we have zero value of x_i, there will be $2s_i$ value in the same row in $\mathbf{X}$. For example, consider the case when $x_1 = 0$, which means $s_1 = -s_3$, because $x_1 = s_1 + s_3$ as defined in (11). This implies that $x_3 = 2s_3$, because $x_3 = s_3 - s_1$, as in (11). We refer to these two elements that have the null value and $2s_i$ as a pair. Investigation of (6) reveals that there are two pairs in each row; that is, the first and third elements form one pair, and the second and fourth elements form the other pair.

We design the AI in order to satisfy the following condition. First, by adding AI, we escape the zero value transmission. For this, whenever we have a zero value of x_i, we add AI that will change the transmitted symbol to a non-zero value. At the same time, we subtract the added value from the paired elements, which reduces the original amplitude. Second, the AI is added in such a way that the paired elements contribute to canceling each other's AI at the receiving end. For this, we multiply the channel gains in such a way that the added AI is canceled.

To achieve the above design goal, we represent the received signal at Bob during the i-th time period, $y_{\mathrm{b}i}$, as follows:

$$y_{\mathrm{b}i} = \sum_{j=1}^{4} \left(\chi_{ij} + \omega_{ij}\right) h_j + n_{\mathrm{b}i}, \tag{17}$$

where, χ_{ij} is the element of $\mathbf{X}$ in the i-th row and j-th column, ω_{ij} is the AI term, and $n_{\mathrm{b}i}$ is the noise at Bob during the i-th time period. Because χ_{ij} and $\chi_{i(j+2)}$, $j = 1, 2$, always form a pair, we represent (17) as the following pairwise form:

$$y_{\mathrm{b}i} = \sum_{j=1}^{2} \left\{ \left(\chi_{ij} + \omega_{ij}\right) h_j + \left(\chi_{i(j+2)} + \omega_{i(j+2)}\right) h_{j+2} \right\} + n_{\mathrm{b}i}. \tag{18}$$

In order to satisfy the first condition, when $\chi_{ij} = 0$, we set ω_{ij} to a non-zero value, so that it can increase the power of χ_{ij}. At the same time, we decrease the same amount of power from the paired element. In other words, we set $\omega_{ij} = \omega_{i(j+2)} = \chi_{i(j+2)}/2$, in the case $\chi_{ij} = 0$. This pairwise power reallocation will contribute to decreasing the difference of power allocations across the transmit antennae. Even though AI is added, it is a kind of reallocation of power, and thus there will be no additional average power allocation. In addition, thanks to the following solution to the second condition, this AI will be perfectly eliminated.

The second condition requires that AI should be canceled at the legitimate receiver. For this, we represent (18) by separating the signal and AI terms, that is

$$y_{\mathrm{b}i} = \sum_{j=1}^{2} \left(\chi_{ij} h_j + \chi_{i(j+2)} h_{j+2}\right) + \sum_{j=1}^{2} \left(\omega_{ij} h_j + \omega_{i(j+2)} h_{j+2}\right) + n_{\mathrm{b}i}. \tag{19}$$

Because the AI term in (19) is $\sum_{j=1}^{2} \left(\omega_{ij} h_j + \omega_{i(j+2)} h_{j+2}\right)$, we need to make it zero, i.e.,

$$\sum_{j=1}^{2} \left(\omega_{ij} h_j + \omega_{i(j+2)} h_{j+2}\right) = 0. \tag{20}$$

A simple solution to (20) can be letting ω_{ij} and $\omega_{i(j+2)}$ contain $-h_{(j+2)}$ and h_j terms, respectively.

By combining the above two conditions at the same time, ω_{ij} can be defined as follows:

$$\begin{array}{lll} \omega_{ij} = \frac{\chi_{i(j+2)}h_{j+2}}{2}, & \omega_{i(j+2)} = -\frac{\chi_{i(j+2)}h_j}{2}, & \text{if } j = 1,2, \text{and } \chi_{ij} = 0, \\ \omega_{ij} = \frac{\chi_{i(j-2)}h_{j-2}}{2}, & \omega_{i(j-2)} = -\frac{\chi_{i(j-2)}h_j}{2}, & \text{if } j = 3,4, \text{and } \chi_{ij} = 0, \\ \omega_{ij} = 0, & & \chi_{ij} \neq 0,\ \chi_{i(j\pm 2)} \neq 0. \end{array} \tag{21}$$

As shown above, ω_{ij} has a non-static time-varying value that is dependent on χ_{ij}, as well as on h_j. At the transmitter part, this AI contributes to reducing the difference of power allocations, while at the receiving end, it will be perfectly canceled out. Therefore, by using (21), (18) can be represented without any AI, as follows:

$$y_{bi} = \sum_{j=1}^{2} \left(\chi_{ij}h_j + \chi_{i(j+2)}h_{j+2}\right) + n_{bi}. \tag{22}$$

In summary, the received signal at Bob during the four time slots can be represented as:

$$\left[y_{b1}\ y_{b2}\ y_{b3}\ y_{b4}\right]^T = (\mathbf{X} + \mathbf{W}) \left[h_1\ h_2\ h_3\ h_4\right]^T + \mathbf{n}_b^T, \tag{23}$$

where $\mathbf{W}$ is the AI matrix with ω_{ij} as the element in the i-th row and j-th column. Then, the received signal after the interference cancelation can be equivalently represented as:

$$\left[y_{b1}\ y_{b2}^*\ y_{b3}\ y_{b4}^*\right]^T = \mathbf{H} \left[s_1\ s_2\ s_3\ s_4\right]^T + \left[n_{b1}\ n_{b2}^*\ n_{b3}\ n_{b4}^*\right]^T, \tag{24}$$

and the detection can be made exactly the same as in the conventional detection using (10).

On the other hand, the received signal vector at Eve is expressed as follows:

$$\left[y_{e1}\ y_{e2}\ y_{e3}\ y_{e4}\right]^T = (\mathbf{X} + \mathbf{W}) \left[g_1\ g_2\ g_3\ g_4\right]^T + \mathbf{n}_e^T, \tag{25}$$

where g_j is the channel gain from the j-th transmit antenna from Alice to Eve. Because the added AI matrix $\mathbf{W}$ has factors related to the channel gains of h_js for Bob, it cannot be canceled out at Eve. That is, the received signal at Eve during the i-th time slot is expressed as follows:

$$y_{ei} = \begin{cases} \sum_{j=1}^{2} \left\{ \left(\chi_{ij} + \frac{\chi_{i(j+2)}h_{j+2}}{2}\right) g_j + \left(\chi_{i(j+2)} - \frac{\chi_{i(j+2)}h_j}{2}\right) g_{j+2} \right\} + n_{ei}, & \text{if } \chi_{ij} = 0, \\ \sum_{j=1}^{2} \left\{ \left(\chi_{ij} - \frac{\chi_{ij}h_{j+2}}{2}\right) g_j + \left(\chi_{i(j+2)} + \frac{\chi_{ij}h_j}{2}\right) g_{j+2} \right\} + n_{ei}, & \text{if } \chi_{i(j+2)} = 0, \\ \sum_{j=1}^{2} \left\{ \chi_{ij}g_j + \chi_{i(j+2)}g_{j+2} \right\} + n_{ei}, & \text{if } \chi_{ij} \neq 0, \chi_{i(j+2)} \neq 0. \end{cases} \tag{26}$$

Even though, there is the probability of not having AI, i.e., $\chi_{ij} \neq 0$ and $\chi_{i(j+2)} \neq 0$, and thus $\omega_{ij} = 0$, frequent uncanceled AI will cause serious performance degradation.

4. Performance Evaluation

This section presents the secrecy and BER performances of the proposed waveform design using LD-QO-STBC over a Rayleigh-faded MIMO channel, and compares them with the conventional schemes. In order to consider the worst-case for secrecy protection, we assume that not only Bob but also Eve have perfect CSI, and that they have the same signal-to-noise ratio (SNR). On the other hand, Alice only shares the CSI for Bob. We also assume that Alice transmits information using 4-QAM through the MIMO channel without frequency selectivity. All channel gains for each transmit symbol are assumed to be independent and constant over four consecutive symbol periods. In the simulations, the aggregated power from all the transmit antennae was normalized to one.

We first compare the secrecy capacity of the proposed scheme with the conventional schemes. It is well known that the secrecy capacity C_s can be expressed by [6]:

$$C_s = \max\left(C_B - C_E, 0\right), \tag{27}$$

where C_B and C_E are the capacities achieved at Bob and Eve, respectively. Because the added AI can be canceled at Bob, C_B can be estimated as in the conventional MIMO systems, as follows:

$$C_B = \log_2 \det\left(\mathbf{I} + \frac{P}{\sigma_b^2} \boldsymbol{H}\boldsymbol{H}^{\mathrm{H}}\right), \tag{28}$$

where P and σ_b^2 are the transmit power and noise variance at Bob, resulting in P/σ_b^2 being the SNR. On the other hand, C_E is estimated by accounting for the added AI, as follows [18]:

$$C_E = \log_2 \det\left(\mathbf{I} + \frac{P}{(\|\mathbf{W}\|^2 + \sigma_e^2)} \mathbf{G}\mathbf{G}^H\right), \tag{29}$$

where, σ_e^2 is the noise variance at Eve.

We estimate the secrecy performance in terms of C_s in (27), and Figure 3 compares the simulation results of the average secrecy capacity. For comparative purpose, we implement a number of conventional schemes: (1) the Alamouti scheme [15], (2) the LD-QO-STBC scheme [14], (3) the QO-STBC scheme with power scaling [13], (4) the Alamouti scheme with AN [10], and (5) the proposed scheme. Because we assume that Bob and Eve have perfect CSI of their own and the same SNR, the secrecy capacities of the conventional schemes without any AN or AI, i.e., (1), (2), and (3), are almost zero. On the other hand, the addition of AN to the Alamouti scheme contributes largely to enhancing the secrecy capacity, even though Eve has the same SNR as Bob. Moreover, the proposed scheme achieves the best secrecy performance amongst the various schemes.

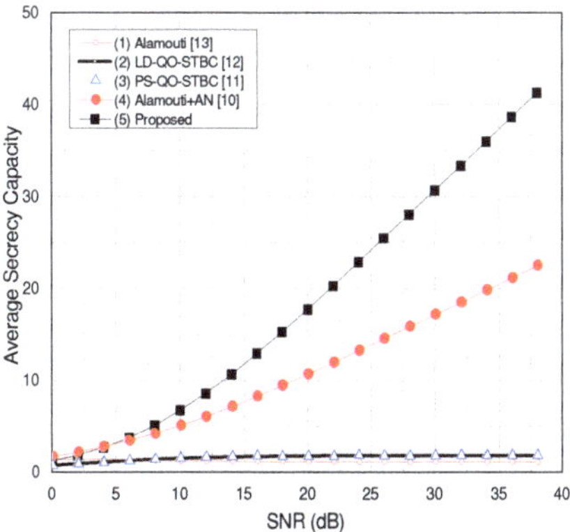

Figure 3. Secrecy capacity of the proposed scheme over a Rayleigh fading channel compared to the conventional space–time block coding (STBC)-based PLS schemes.

Next, Figure 4 compares the BER performance simulation results. With the conventional schemes, (1), (2), and (3), Bob and Eve will have almost the same BER performance. On the other hand, due to the highly enhanced secrecy protection in (4) the Alamouti scheme with AN, and (5) the proposed scheme, the BER performance at Eve approaches to 0.5. This means that there is no information leakage to Eve. However, at the expense of power usage of adding AN in scheme (4), we have to sacrifice power loss. For example, in scheme (4) the same power is allocated to AN as to the information signal, so we can find a 3 dB power loss in the performance. On the other hand, the proposed scheme utilizes the pairwise power reallocation for AI, and thus it does not incur any power loss.

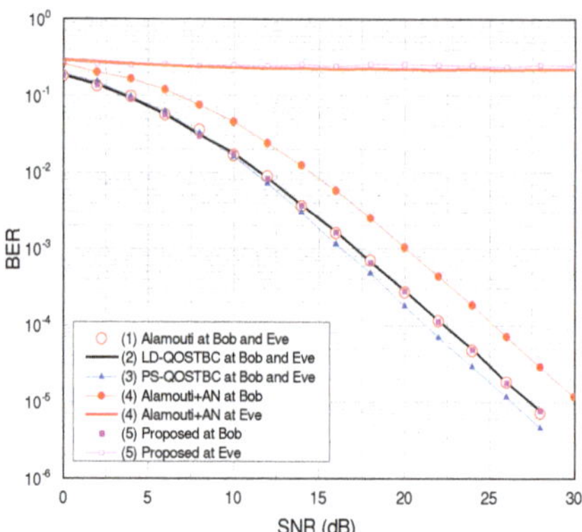

Figure 4. Comparison of bit error rate (BER) performances for various STBC-based PLS schemes over a Rayleigh fading channel at the legitimate (Bob) and illegal (Eve) receivers.

5. Conclusions

In this paper, we presented a novel secrecy-enhancing scheme by tailoring the waveform of the LD-QO-STBC scheme. The proposed scheme employed dynamic AI that was dependent on the signal, as well as the channel gain of the legitimate receiver. The dependency on the information signal was used to reduce the difference of power allocations across the transmit antennae. In addition, the dependency on the channel gain was used to cancel the AI only at the legitimate receiver, while imposing serious interference on the passive eavesdropper. As a result, the secrecy capacity was highly enhanced. The BER simulation results showed that the proposed scheme prevented information leakage to the illegal listener, even in the case when the illegal listener has a perfect CSI and the same SNR as the legitimate receiver. Because the proposed method does not incur any power loss due to the addition of AI, it can be efficiently utilized for many wireless systems using STBC, resulting in diversity gain as well as security protection. Based on the promising results of the proposed scheme, our future work can be directed to the investigation of dynamic waveform design and efficient detection schemes utilizing artificial intelligence techniques [19,20].

Author Contributions: Conceptualization, S.K. and H.L.; methodology, P.S.; software, P.S.; validation, P.S., H.L. and S.K.; writing–original draft preparation, P.S.; writing–review and editing, S.K.; supervision, S.K.; project administration, S.K.; funding acquisition, S.K. All authors have read and agreed to the published version of the manuscript.

Funding: The research was supported by Basic Science Research Program through the National Research Foundation of Korea (NRF) funded by the Ministry of Education (2017R1D1A1B03027939).

Conflicts of Interest: The authors declare no conflict of interest.

References

1. Coskun, A.F.; Kucur, O. Secrecy outage probability of conventional and modified TAS/Alamouti-STBC schemes with power allocation in the presence of feedback errors. *IEEE Trans. Veh. Technol.* **2019**, *68*, 2609–2623. [CrossRef]

2. Hamamreh, J.M.; Furqan, H.M.; Arslan, H. Classifications and applications of physical layer security techniques for confidentiality: A comprehensive survey. *IEEE Commun. Surv. Tutor.* **2018**, *21*, 1. [CrossRef]
3. Wu, Y.; Khisti, A.; Xiao, C.; Caire, G.; Wong, K.-K.; Gao, X. A survey of physical layer security techniques for 5G wireless networks and challenges ahead. *IEEE J. Sel. Areas Commun.* **2018**, *36*, 679–695. [CrossRef]
4. Yener, A.; Ulukus, S. Wireless physical-layer security: Lessons learned from information theory. *Proc. IEEE* **2015**, *103*, 1814–1825. [CrossRef]
5. Liu, T.; Shamai, S. A note on secrecy capacity of multiple-antenna wiretap channel. *IEEE Trans. Inf. Theory* **2009**, *55*, 2547–2553. [CrossRef]
6. Oggier, F.; Hassbii, B. The secrecy capacity of the MIMO wiretap channel. *IEEE Trans. Inf. Theory* **2011**, *57*, 4961–4972. [CrossRef]
7. Shang, P.; Yu, W.; Zhang, K.; Jiang, X.-Q.; Kim, S. Secrecy enhancing scheme for spatial modulation using antenna selection and artificial noise. *Entropy* **2019**, *21*, 626. [CrossRef]
8. Yan, S.; Yang, N.; Malaney, R.; Yuan, J. Transmit antenna selection with Alamouti coding and power allocation in MIMO wiretap channels. *IEEE Trans. Wirel. Commun.* **2014**, *13*, 1656–1667. [CrossRef]
9. Allen, T.; Tajer, A.; AL-Dhahir, N. Secure Alamouti MAC transmissions. *IEEE Trans. Wirel. Commun.* **2017**, *16*, 3674–3682. [CrossRef]
10. Shang, P.; Kim, S.; Jiang, X.-Q. Efficient Alamouti-coded spatial modulation for secrecy enhancing. In Proceedings of the 2019 International Conference on Information and Communication Technology Convergence (ICTC), Jeju, Korea, 16–18 October 2019.
11. Pouri, A.B.; Torabi, M. Physical layer security in space-time block codes from coordinate interleaved orthogonal design. *IET Commun.* **2020**, *14*, 2007–2017. [CrossRef]
12. Li, M.L.; Yuan, H.; Yue, X.; muhaidat, S.; Maple, C.; Dianati, M. Secrecy outage analysis for Alamouti space–time block coded Non-Orthogonal multiple access. *IEEE Commun. Lett.* **2020**, *24*, 1405–1409. [CrossRef]
13. Chae, S.; Bang, I.; Lee, H. Physical layer security of QSTBC with power scaling in MIMO wiretap channels. *IEEE Trans. Veh. Technol.* **2020**, *69*, 5647–5651. [CrossRef]
14. Park, U.; Kim, S.; Lim, K.; Li, J. A novel QO-STBC scheme with linear decoding for three and four transmit antennas. *IEEE Commun. Lett.* **2008**, *12*, 868–870. [CrossRef]
15. Alamouti, S.M. A simple transmit diversity technique for wireless communications. *IEEE J. Sel. Areas Commun.* **1998**, *16*, 1451–1458. [CrossRef]
16. Jafarkhani, H. A quasi-orthogonal space-time block code. In Proceedings of the 2000 IEEE Wireless Communications and Networking Conference, Chicago, IL, USA, 23–28 September 2000.
17. Tirkkonen, O.; Boariu, A.; Hottinen, A. Minimal non-orthogonality rate 1 space-time block code for 3+ Tx antennas. In Proceeding of the 2000 IEEE Sixth International Symposium On Spread Spectrum Techniques and Applications, Parsippany, NJ, USA, 6–8 September 2000.
18. Negi, R.; Goel, S. Secret communication using artificial noise. In Proceedings of the 2005 IEEE 62nd Vehicular Technology Conference, Dallas, TX, USA, 28–28 September 2005.
19. Martino, L.; Read, J. Joint Introduction to Gaussian Processes and Relevance Vector Machines with Connections to Kalman Filtering and other Kernel Smoothers. Available online: https://arxiv.org/abs/2009.09217 (accessed on 19 September 2020).
20. O'Shea, T.J.; Erpek, T.; Clancy, T.C. Deep Learning Based MIMO Communications. Available online: https://arxiv.org/abs/1707.07980 (accessed on 25 July 2017).

Publisher's Note: MDPI stays neutral with regard to jurisdictional claims in published maps and institutional affiliations.

Article

Performance Analysis of LDS Multi Access Technique and New 5G Waveforms for V2X Communication

Imane Khelouani [1,2,*], Fouzia Elbahhar [1], Raja Elassali [2] and Noureddine Idboufker [2]

1 COSYS-LEOST, University Gustave Eiffel, 59650 Villeneuve d'ascq, France; fouzia.boukour@univ-eiffel.fr
2 TIM, ENSA, University of Cadi Ayyad, Marrakech 40000, Morocco; r.elassali@uca.ma (R.E.); n_idboufker@yahoo.fr (N.I.)
* Correspondence: imane.khelouani@univ-eiffel.fr

Received: 1 May 2020; Accepted: 2 July 2020; Published: 4 July 2020

Abstract: Low Density Signature (LDS) is an emerging non-orthogonal multiple access (NOMA) technique that has never been evaluated under a vehicular channel in order to simulate the environment of a vehicle to everything (V2X) communication. Moreover, the LDS structure has been combined with only Orthogonal Frequency Division Multiplexing (OFDM) and Filter-Bank Multi-Carrier (FBMC) waveforms to improve its performances. In this paper, we propose new schemes where the LDS structure is combined with Universal Filtered Multi Carrier (UFMC) and Filtered-OFDM waveforms and the Bit Error Rate (BER) is analysed over a frequency selective channel as a reference and over a vehicular channel to analyse the effect of the Doppler shift on the overall performance.

Keywords: V2X; LDS-F-OFDM; LDS-UFMC; EVA channel model

1. Introduction

Vehicular communications have recently caught a lot of attention in the research community thanks to the advantages that they can provide the overall vehicular experience. In fact, several technologies have provided the requirements for this type of communication. Connected to the infrastructure (V2I), to another vehicle (V2V) or to a pedestrian (V2P), the vehicle to everything (V2X), as it is referred to, is a new solution for road users to enhance safety and improve the traffic efficiency. The Long Term Evolution (LTE) as a widely deployed infrastructure is proposed to be extended in order to support the V2X services, namely the LTE-based V2X [1]. The 3rd Generation Partnership Project (3GPP) Release 14 is an evolutionary standard that is dedicated to the LTE-based V2X and published on September 2016 defining two new modes, mode 3 and mode 4 [2,3]. In mode 3, the radio resource is managed by the cellular network, consequently vehicles can only communicate under a cellular coverage. Meanwhile in mode 4, the radio resource is managed autonomously by the vehicle itself for the direct V2V communication overcoming the coverage limitation of mode 3. However, LTE-based V2X standard suffers from severe performance degradation in a high density environment—it does not allow a high number of users access to the network. In addition, LTE-based V2X is based on Single Carrier Frequency Division Multiple Access (SC-FDMA) which requires high complexity equalizers. The ITS-G5 is another standard that is introduced by the European Telecommunications Standards Institute (ETSI) and operates in 5 GHz frequency band [4]. The main advantage of ITS-G5 is its low latency, the short transmission delay is due to the fact that data is being transferred directly between neighbours. However, in best case scenarios, ITS-G5 has a short range of 1 km and is extremely sensitive to dense environment which reduces the total throughput and increases the end-to-end latency. Similar to the IEEE 802.11p US standard, both LTE-based V2X and ITS-G5 rely on Orthogonal Frequency Division Multiplexing (OFDM) on the PHY layer, while maintaining the subcarriers orthogonality could be challenging in the vehicular environment leading to a high Bit Error Rate (BER). This is why the 5G cellular network can be a promising technology to support the vehicular communications [5], namely New Radio-V2X (NR-V2X).

The 3GPP Release 16 defines the first specifications for NR-V2X sidelink where it supports subcarrier spacings of 15 kHz, 30 kHz, 60 kHz and 120 kHz. Their associations to Cyclic Prefix (CPs) and frequency ranges are as for NR Uplink/Downlink (UL/DL), but using only the CP-OFDM waveform. The modulation schemes available are Quadrature Phase Shift Keying (QPSK), Quadrature Amplitude Modulation 16-(QAM), 64-QAM and 256-QAM [6]. Meanwhile, another study has been published for non-orthogonal multiple access (NOMA) signature candidates [7] proposing some of the main technologies such as Interleaver Division Multiple Access (IDMA) [8–10], Pattern Division Multiple Access (PDMA) [11,12] and Sparse Code Multiple Access (SCMA) [13,14]. However, for V2X communication, no specifications invoke NOMA.

As an efficient multiple access technique, we have chosen to modulate Low Density Signature (LDS) and analyse its performances under a vehicular channel for V2X communication. Its sparse structure enables each user to spread its data over a small subset of subcarriers, which means that a single subcarrier will support only a small number of users or symbols, hence reducing the multiuser interference (MUI). Firstly, it is proposed in [15] for a Code Division Multiple Access (CDMA) system and proves that it can afford a high system overload with affordable complexity. To improve its performances, LDS is enhanced by combining it with OFDM [16] to apply the spreading over OFDM subcarriers and evaluate over a frequency selective channel. Furthermore, the LDS has been recently combined with FBMC [17] and a joint sparse graph receiver combing pulse shaping property, NOMA and channel coding is proposed to improve the overall result at the cost of very high complexity. Although LDS-OFDM and Joint Sparse Graph-Isotropic Orthogonal Transfer Algorithm (JSG-IOTA) have shown improved performance evaluation, these schemes have only been analysed over a frequency selective channel. Our objective is to extend the state-of-art work to evaluate the performance of LDS over a high mobility channel to simulate the vehicular environment for V2X applications and combine it with other advanced 5G waveforms, specifically for an application that requires a high data rate, a medium number of users connected to the network in a certain geographical area and a time of latency smaller than 1 ms [5].

Among these promising 5G waveforms, Universal Filtered Multi-Carrier (UFMC) [18] has drawn attention due to its ability to overcome OFDM shortcomings. By applying properly designed sub-band filtering, UFMC reduces the high out-of-band (OOB) power emission while retaining the simplicity of the conventional OFDM signal. In fact, one of the main advantages of UFMC is its compatibility with the OFDM signals which achieves low system complexity in the NOMA schemes.

Filtered-OFDM [19,20] is also one of the proposed advanced waveforms for the future cellular network. Both UFMC and f-OFDM signals use filtering per subband in order to achieve a low OOB, the main difference is the filter length and its flexibility. F-OFDM uses a long filter with different lengths for each subband (Windowed Sinc filter) exceeding the CP while UFMC signals use short fixed length filters for each subband (Chebyshev filter). In this paper, our goal is to propose a scheme that benefits from the waveforms robustness and multi-carrier transmission combined with the non-orthogonality of a spreading based NOMA to provide a system with less complexity while keeping the transmission model flexible to any future changes and maintaining a certain compatibility with the current techniques. It is why, we have developed a new LDS-UFMC and LDS-F-OFDM schemes in which the LDS spreading is applied to UFMC and f-OFDM signals. Among the different filters that f-OFDM can offer, we adopt computing the Hann, Hamming and Blackman filters thanks to the great performance they shown in terms of BER and OOB reduction [21] and also to provide a better comparison. These schemes will be BER analysed for a V2X communication i.e., a high mobility channel. The multiuser detector will be based on Message Passing Algorithm (MPA) [22]. It will be proved that LDS-F-OFDM outperforms the LDS-OFDM and LDS-UFMC by allowing the filter length to exceed the CP length of OFDM and designing the filter appropriately.

The remainder of this paper is organised as follows. Section 2 introduces the system model of LDS-UFMC. Section 3 is devoted to the LDS-F-OFDM system model. Section 4 presents simulation results of LDS-UFMC and LDS-F-OFDM compared to LDS-OFDM in different environments. Finally, Section 5 concludes this work.

2. LDS-UFMC System Model

In this section, we are going to define the major blocks of an LDS-UFMC system. The LDS-UFMC block diagram is shown in Figure 1. After coding and modulation, the symbols are multiplied with low density spreading sequences then modulated and transmitted simultaneously. Afterwards, spreaded data are shaped into a time-frequency grid. In this configuration, we address the pilots aided channel estimation. For simplicity, we propose to insert the pilots using the Comb Type arrangement. Then, this grid is UFMC modulated using an N-point Inverse Fast Fourier Transform (IFFT) and filtered by a L-length Dolph–Chebyshev filter. At the receiver side, the received signal is UFMC demodulated, then the pilots extraction is performed in order to estimate the channel using the Least Square (LS) method and linear interpolation. The equalizer block uses Zero-Forcing to eliminate the channel effect. The output of the equalizer is passed to the LDS detector where an iterative detection process based on MPA is performed to separate the users' symbols.

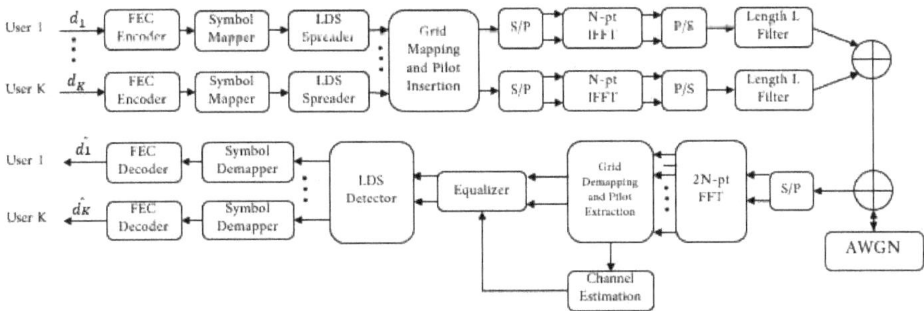

Figure 1. Low Density Signature-Universal Filtered Multi-Carrier (LDS-UFMC) block diagram.

2.1. LDS Spreader

Consider an LDS-UFMC system with K users and the users are indexed as follows; $k = 1, ..., K$ and all users are assumed to transmit the same number of data streams M and they are indexed as follows; $m = 1, ..., M$. Assume the number of subcarriers is N, and they are indexed as follows; $n = 1, ..., N$. We define the spreading matrix of the kth user as:

$$\mathbf{S_k} = [\mathbf{s_{k,1}}, ..., \mathbf{s_{k,M}}] \in \mathbb{C}^{N \times M} \tag{1}$$

$\mathbf{s_{k,m}} = [s^1_{k,m}, ..., s^N_{k,m}]^T$ is the sparse vector of length N used to spread the mth symbol of the kth user. Thus, the matrix of spreading of all users can be represented as :

$$\mathbf{S} = [\mathbf{S_1}, ..., \mathbf{S_K}] \in \mathbb{C}^{N \times MK} \tag{2}$$

Thanks to the sparse nature of this matrix, only a small number of users can share the same subcarrier, we define it as d_c the interference degree. Hence, among N subcarriers only d_v will be used to serve one user, d_v is called the effective spreading gain. Unlike the conventional CDMA system, we require $d_v \ll N$ and $d_c \ll K$. Let $\mathbf{a_k}$ be user's k symbols:

$$\mathbf{a_k} = [a_{k,1}, ..., a_{k,M}]^T \tag{3}$$

After the spreading process, the signal of the kth user $\boldsymbol{x_k} = [x_k^1, ..., x_k^N]^T$ is a vector of length N and can be written as follows:

$$\boldsymbol{x_k} = \boldsymbol{S_k} \times \boldsymbol{a_k} \tag{4}$$

and x_k^n will be the data transmitted over the nth subcarrier by the kth user:

$$x_k^n = \sum_{m=1}^{M} a_{k,m} s_{k,m}^n \tag{5}$$

Hence, the signal transmitted over the nth subcarrier is given by (6):

$$x^n = \sum_{k \in \xi_n} x_k^n \tag{6}$$

ξ_n is considered as the group of users interfering in the nth subcarrier.

Figure 2 represents the spreading process of the LDS. In this example, we consider that each user transmits one symbol (i.e., M = 1). The system transmits over 4 subcarriers and serves 8 users which means that the overloading is at 200%. Each subcarrier is allocated to 4 users (d_c = 4), and each user spreads its data over 2 subcarriers (d_v = 2). The symbols are the variable nodes and the subcarriers are the functions nodes respectively. The edges that connect the nodes define if a symbol will be spreaded over the adjacent subcarrier. For instance, the first user is connected to subcarriers 1 and 2, consequently, the first symbol will be spreaded over these adjacent subcarriers only. This representation of the spreading process is called the Tanner Graph [16].

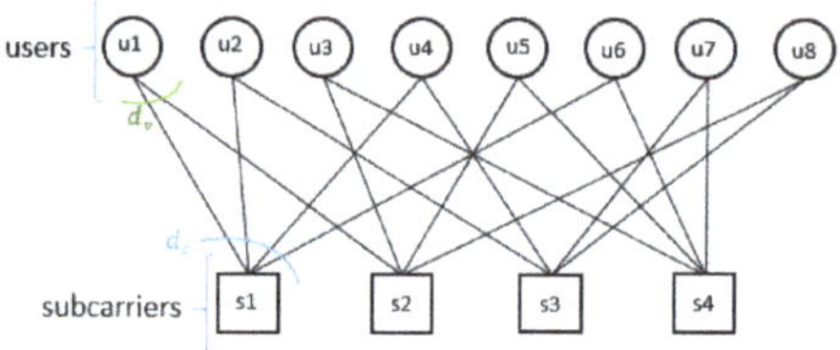

Figure 2. Spreading process of LDS-UFMC system.

2.2. Signal Model

After the LDS spreading, the UFMC modulation is applied to each subcarrier. In fact, UFMC divides the entire bandwidth B of N subcarriers into multiple subbands S_u, each subband consists of Q subcarriers. Hence, the nth subcarrier can now be regarded as the qth subcarrier of the sth subband. After that, the IFFT is performed over the signal of each subband and will be filtered using a L-length prototype filter, usually, the filter is a Dolph–Chebyshev window. The subbands are then superimposed and sent over the channel. The baseband discrete UFMC signal is represented in (7):

$$x_{LDS-UFMC}[\boldsymbol{n}] = \sum_{s=0}^{S_u-1} g_s[\boldsymbol{n}] \times x_s[\boldsymbol{n}] \tag{7}$$

$g_s[\boldsymbol{n}]$ defines the filter used in the sth subband :

$$g_s[\boldsymbol{n}] = g[n] e^{j(2\pi n Q/2)/N} e^{j(2\pi n(S_0 + sQ))/N} \tag{8}$$

where S_0 denotes the starting frequency of the lowest subband and $g[n]$ is a well-localised Chebyshev pulse filter of length L. The third term of the equation is the one responsible for frequency shifting to the appropriate subband. $x_s[n]$ is signal transmitted over the sth group of subcarriers. It can be expressed as follows:

$$x_s[n] = \sum_{q=0}^{Q-1} x^{s,q} e^{j(2\pi nq/2)/N} e^{j(2\pi n(S_0+sQ))/N} \tag{9}$$

Hence, $x^{s,q}$ can be nothing but the signal transmitted over the qth subcarrier of the sth subband and just like in (6), it is given by:

$$x^{s,q} = \sum_{k \in \xi_{s,q}} x_k^{s,q} \tag{10}$$

where $\xi_{s,q}$ is considered as the group of users interfering in the qth subcarrier of the sth subband.

The received signal is represented as follows:

$$y_{LDS-UFMC}[n] = h[n] \times x_{LDS-UFMC}[n] + z[n] \tag{11}$$

where $z[n]$ and $h[n]$ are the additive white gaussian noise with variance σ_z^2 and the channel impulse response respectively. The length of the signal $y_{LDS-UFMC}$ is $N_y = N + L - 1$ due to the convolution with the subband filter, consequently, a 2N-point Fast Fourier Transform (FFT) is performed at the UFMC receiver. After the UFMC demodulation process, the input of the LDS Detector corresponding to the nth subcarrier is :

$$Y_n = H_n \sum_{k \in \xi_n} \sum_{m=1}^{M} a_{k,m} s_{k,m}^{n} \tag{12}$$

where Y_n and H_n are the 2N-point FFT of the time domain signal respectively. In our system, we have chosen to use the equalizer embedded with the LDS detector for a better performance. Regarding the LDS decoding, firstly, the LDS turbo receiver uses MPA and Forward Error Correction (FEC) decoder to find the reliability of the symbols. Secondly, the JSG receiver uses pulse shaping, NOMA and channel coding, however, it is highly complex. For simplicity, we chose to implement the basic LDS detector that uses the Tanner Graph for implementing the MPA receiver in which we consider subcarriers and symbols as function nodes and variable nodes, respectively. Adjacent nodes are connected via edges. Based on an extrinsic manner, each node will update its information containing the reliability of the symbol based on the received reliability from other edges and send it back. After an appropriate number of iterations, the reliability which is the log likelihood ratio (LLR) of the symbols will converge and the symbols are transmitted to the FEC decoder. The major goal behind this complex implementation is to find the value of $\hat{a}$ that maximises the joint a posteriori probability based on the observed signal:

$$\hat{a} = \arg\max_{a \in \mathbb{X}} p(a|y) \tag{13}$$

where $\mathbb{X}$ is the modulation constellation. The first LLR update can be written as follows:

$$\begin{aligned} \ell_{c_n \leftarrow u_k}(a_k) &= \log \frac{P_{\text{ext},n}(a_k = +1)}{P_{\text{ext},n}(a_k = -1)} \\ &= \sum_{m \in \varphi_k \backslash n} \ell_{c_m \rightarrow u_k} \end{aligned}$$

where u nodes are the variable nodes and c nodes are the function nodes, hence $\ell_{c_n \leftarrow u_k}$ is the message sent from u_k node to the c_n node and $\ell_{c_n \rightarrow u_k}$ message sent from c_n node to the u_k node, respectively. It is clear from the above equation that the update of $\ell_{c_n \leftarrow u_k}(a_k)$ is dependent to all nodes besides the node n i.e., $m \in \varphi_k \setminus n$, hence, the notation of extrinsic update. Second Update can be calculated as follow:

$$\ell_{c_n \rightarrow u_k}(a_k) = \log\Big(\sum_{a_n \in X} P(y_n|a_n) \prod_{l \in \xi_n \backslash k} P_n(a_l) \Big)$$

where φ_k is the group of subcarriers allocated to the kth user and $P(y_n|a_n)$ is the channel observation function at subcarrier y_n. Further explanations are provided in [22].

3. LDS-F-OFDM System Model

Based on the same LDS spreader used in LDS-UFMC, LDS-F-OFDM applies Filtered-OFDM waveform on the users' spreaded signals and transmits them over the channel. The LDS based on f-OFDM block diagram is presented in Figure 3. As depicted, the signal is mapped into a time-frequency grid and then the pilots are inserted for channel estimation. Afterwards, a N-point IFFT is performed to transfer the grid from the frequency domain to the time domain, data is then serialised and CP is added to combat Inter Symbol Interference (ISI) effect. Like UFMC, the main advantage of f-OFDM is its compatibility with OFDM, with a difference of passing the signal through a long length filter before the transmission. Extended over the CP, the filter helps to lower the high OOB radiation while maintaining the OFDM orthogonality between the subcarriers. Usually, a truncated sinc window is used as a shaping filter, however in our system we have chosen to implement other types of filters for a better comparison. For LDS-F-OFDM, we consider all the subcarriers as one subband to lower the overall complexity. Therefore, our transmitted signal is an OFDM modulated signal passed through an appropriate shaped filter. Without loss of generality, the signal expression can be written as follows:

$$x_{LDS-F-OFDM}[n] = x_{LDS-OFDM}[n] \times f_{TX}[n] \tag{14}$$

where $f_{TX}(n)$ represents the well-designed filter and $x_{LDS-OFDM}[n]$ is the discrete LDS-OFDM signal after the spreading and modulation, it is represented as follows:

$$x_{LDS-OFDM}[n] = \sum_{q=0}^{N-1} x^q e^{j2\pi nq/N} \tag{15}$$

The transmitted signal is given by (16):

$$x_{LDS-F-OFDM}[n] = \sum_{q=0}^{N-1} x^q e^{j2\pi nq/N} \times f_{TX}[n] \tag{16}$$

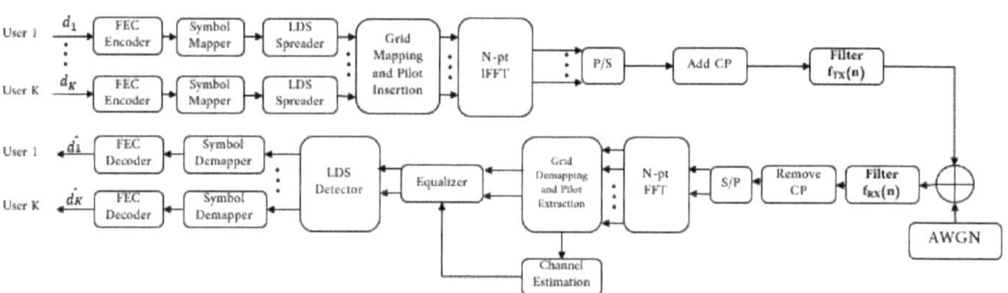

Figure 3. LDS-F-OFDM block diagram.

3.1. Filter Design

The main reason behind the filtering is to eliminate the side lobes for an efficient transmission, it is why filters based on cardinal sin are taken into consideration thanks to their sharp filter response. However, the filter characteristic should be truncated due to its infinite impulse response. As a result, we obtain a finite sinc filter and a window pulse w[n] is applied to smooth its transitions. Once the filter is computed, it should be shifted around the central frequency of the subband.

$$p[n] = sinc(\frac{W + 2\delta_W}{N} n) \tag{17}$$

$$p_B[n] = \frac{p[n]w[n]}{\sum_n |p[n]w[n]|} \quad (18)$$

$$f_{TX}[n] = p_B[n]e^{\frac{j2\pi n f_c}{N\Delta_f}} \quad (19)$$

where W is the number of active subcarriers, δ_W is the tone offset, N is the FFT size, f_c is the frequency of the centre subcarrier in the baseband and Δ_f is the subcarrier spacing. The window function of the chosen windows are defined in Table 1.

Table 1. Window function of the filters.

Window	Window Function
Hann	$w[n] = 0.5 \times (1 + cos(\frac{2\pi n}{L-1}))$
Hamming	$w[n] = \frac{25}{46} + \frac{21}{45} \times cos(\frac{2\pi n}{L-1})$
Blackman	$w[n] = \frac{7938}{18608} + \frac{9240}{18608} \times cos(\frac{2\pi n}{L-1}) + \frac{1430}{18608} \times cos(\frac{4\pi n}{L-1})$

At the receiver side, a matching filter $f_{RX}[n]$ to the transmission filter $f_{TX}[n]$ is performed at the received signal, after that the CP is removed and the signal is transmitted to an OFDM receiver. After the grid demapping and pilot extraction, channel estimation is performed based on the pilots. The output data are then considered as the input of LDS detector explained in Section 2.

4. Performance Evaluation

In this section, we propose to compare the performances of the proposed LDS-F-OFDM and LDS-UFMC schemes with the LDS-OFDM over a vehicular channel with different speed limits and over a multipath fading channel. The main simulation parameters are used to match the 3GPP Release 16 standard. Based on 3GPP Release 16, we have fixed the total bandwidth to $B = 10$ MHz and a carrier frequency of $f_c = 5.9$ GHz to match the NR-V2X operating bands in FR1 [23]. Table 2 summarises the simulation parameters.

Table 2. Simulation parameters.

Parameters	Symbol	Value
Release 16 Parameters		
Bandwidth	B	10 MHz
Carrier frequency	f_c	5.9 GHz
Number of symbols	N_{sym}	14
Subcarriers spacing	Δ_f	15 kHz 30 kHz
Number of Resource Blocks	N_{RB}	52 24
FFT Size	N_{FFT}	1024 512
Cyclic Prefix Length	L_{CP}	72 36

Table 2. *Cont.*

Parameters	Symbol	Value
Pilots Parameters		
Pilots Arrangement	-	Comb type
Pilots Spacing	N_{ps}	4
Number of Pilots per symbol	N_P	156 72
UFMC		
Filter Length	L_U	65 33
Filter Type	-	Dolph-Chebyshev
Side Lobe Attenuation	-	50 dB
Number of Subbands	S	52 24
Subband size	Q	12
F-OFDM		
Filter Length	L_F	513 257
Window function	-	Hann Hamming Blackman
LDS Scheme Parameters		
Effective Processing gain	d_v	3
Interference Pattern	d_c	3
Number of Users	N_u	468 216
Modulation	-	BPSK

No error correction is employed, so the evaluation is carried out for uncoded bits. The LDS spreading matrix is generated randomly for all the simulated schemes. However, we do believe that a well designed matrix with small number of cycle-of-four and a high girth can achieve better performance.

In order to respect the 3GPP recommendations, no overloading scenario is deployed due to the use of pilots and the waveforms offset. We have overcome this circumstance by respecting the rule of interfering several users in the same resource and allocating a few subcarriers to the same user i.e., respecting the LDS structure to justify these new schemes performances. The maximum number of iterations of the LDS detector is ten.

The first channel we have chosen to evaluate the performance of these schemes is the Tapped Delay Line model, specifically the TDL-A channel. The TDL-A channel model has a Doppler spectrum which is characterised by a Jake's spectrum shape. The Power Delay Profile (PDP) of the model is presented in [24] and the delay spread used to scale the normalised taps delays is $D_s = 93$ ηs. This delay spread is chosen to correspond to a short delay profile in an Urban Macro environment for a 5.9 GHz carrier frequency.

Figure 4 depicts the simulation results of LDS-F-OFDM and LDS-UFMC over TDL-A channel. We first fix the bandwidth at $B = 10$ MHz, then we investigate the schemes under two systems, the first with 1024 subcarriers and $\Delta_f = 15$ kHz subcarrier spacing and the second with 512 subcarriers and $\Delta_f = 30$ kHz. According to the simulation results, the LDS-F-OFDM shows better performance over both LDS-UFMC and LDS-OFDM. The fact that the f-OFDM subband regroups all the available subcarriers concentrates the energy in one main lobe. This results in less ISI, and thanks to the LDS structure, the MUI are eliminated, thus, reducing the overall interference. Hence, it is expected that the f-OFDM with the LDS structure achieve a better BER performance. Meanwhile, LDS-UFMC shows a slight performance degradation compared to LDS-OFDM. However, UFMC offers lower OOB and the main advantage of UFMC is the increased spectral efficiency. Consequently, a trade-off must be taken in consideration depending on the requirements of the application.

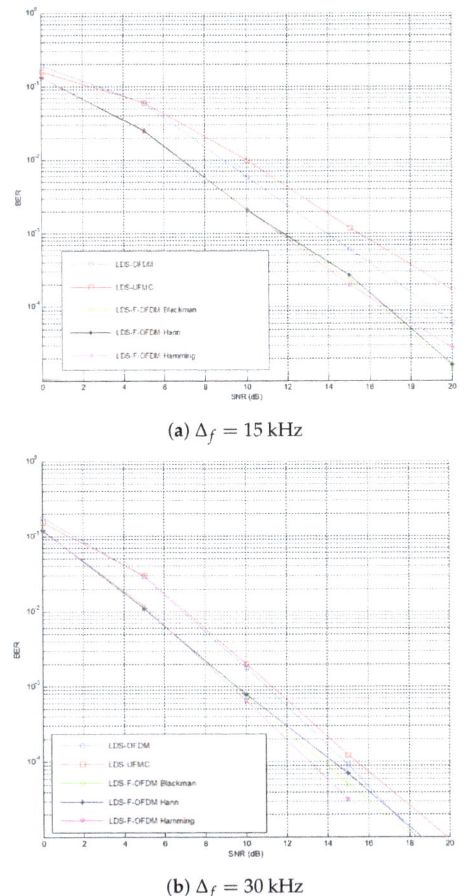

(**a**) $\Delta_f = 15$ kHz

(**b**) $\Delta_f = 30$ kHz

Figure 4. Performance of the proposed schemes over a Tapped Delay Line (TDL)-A channel.

The second and main channel used to simulate these schemes is the Extended Vehicular A (EVA) model channel, its power delay profile is provided in [25]. Similarly to the TDL-A model, the EVA model is characterised by a Jake's Doppler spectrum and the maximum speed simulated is 300 km/h, hence the maximum Doppler shift $f_d = 556$ Hz and specifically modelled to simulate a vehicular channel.

Figure 5a shows the BER performance of the LDS-F-OFDM and LDS-UFMC compared to LDS-OFDM over an EVA channel with a computed speed of 100 km/h which means that the maximum Doppler shift is $f_d = 185$ Hz with a subcarrier spacing of $\Delta_f = 15$ kHz and an FFT size of 1024, hence only 468 subcarriers left for data transmissions. For simplicity, we consider 468 users each sending one BPSK modulated symbol with a power of 1 Watt. In fact, this is the worst loading scenario where we have the number of users equal to the number of subcarriers allocated to data transmission which means that each user will be able to send only one symbol. It can be noticed that the LDS-F-OFDM slightly outperforms once more both LDS-OFDM and LDS-UFMC with its all different types of filters deployed. While Hann, Hamming and Blackman windows achieve almost the same BER performance, Hann window function is the filter susceptible to give better OOB performance. In terms of spectral efficiency, LDS-UFMC compared to LDS-F-OFDM and LDS-OFDM comes first due to the removal of the CP. However, observing the used subcarriers for data transmission due to the use of the pilot with only 4 subcarrier spacing, we notice the overall degraded spectral efficiency. This is completely justified by the need of an efficient channel estimation because of the high changing nature of the channel.

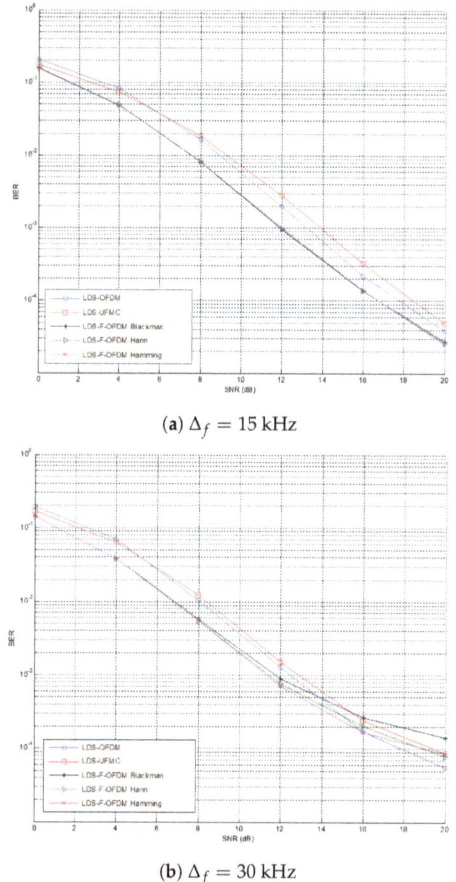

(**a**) $\Delta_f = 15$ kHz

(**b**) $\Delta_f = 30$ kHz

Figure 5. Performance of the proposed schemes over an Extended Vehicular A (EVA) channel with speed of 100 km/h.

The BER performance of the proposed schemes with a computed speed of 100 km/h and a subcarrier spacing of $\Delta_f = 30$ kHz and 512 subcarriers is presented in Figure 5b. It is worth noting that LDS-F-OFDM shows again better performance at low SNR, while at higher SNR, all the proposed schemes face slight performance degradation. In fact, while employing higher subcarrier spacing to combat the Doppler shift, the spectral efficiency is reduced and also the number of subcarriers allocated to pilots is reduced (156 subcarriers in the first system vs. 72 in the second). Consequently, the channel estimation will not be as efficient as in the first system. Furthermore, in order to investigate how the LDS-F-OFDM and LDS-UFMC can be affected by the mobility, the schemes are also simulated over an EVA channel with a speed of 300 km/h.

Figure 6a,b represents the simulation results for $\Delta_f = 15$ kHz and $\Delta_f = 30$ kHz, respectively. It can be seen that LDS-F-OFDM shows better performances than LDS-OFDM and LDS-UFMC in low SNR. Moreover, all the techniques demonstrate huge improvements in the second system with $\Delta_f = 30$ kHz subcarrier spacing than the first. Thanks to the large subcarrier spacing, the Doppler shift does not impact severely the subcarriers orthogonality. In addition, comparing Figure 5a with Figure 6a, all the techniques face severe performance degradation due to the high mobility of the UEs. Meanwhile, Figure 5b with Figure 6b presents almost the same BER performances thanks to the large subcarrier spacing.

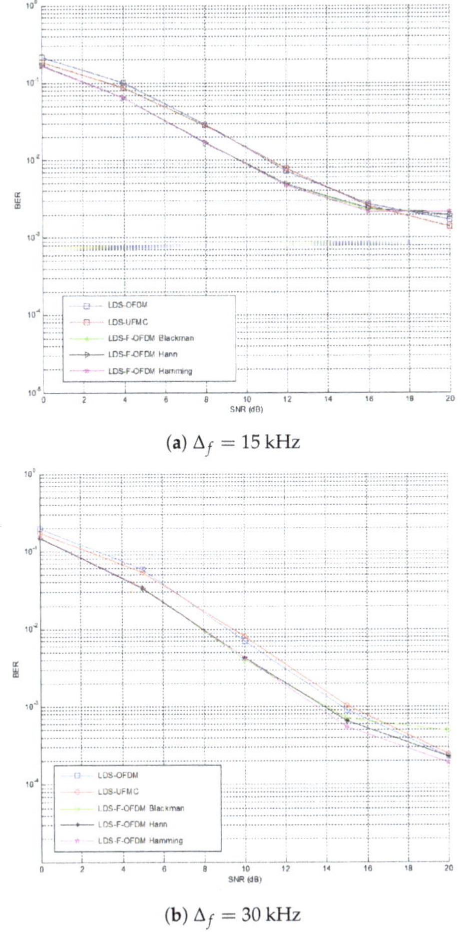

(**a**) $\Delta_f = 15$ kHz

(**b**) $\Delta_f = 30$ kHz

Figure 6. Performance of the proposed schemes over an EVA channel with speed of 300 km/h.

Figure 7 depicts the comparison of the proposed schemes over the EVA channel with different speed limits with both systems. We have chosen to evaluate the schemes also over the EVA channel with a 500 km/h velocity and analyse the degradation of the performance in such an environment. Obviously, the performance decreases due to the high speed, however as shown in Figure 7b i.e., with a large subcarrier spacing, both LDS-F-OFDM and LDS-UFMC seem to suffers less. Meanwhile with $\Delta_f = 15$ kHz, all the techniques seem unable to converge at high speed. To improve this, other receivers can be considered such as the JSG and Expectation Propagation Algorithm EPA receivers.

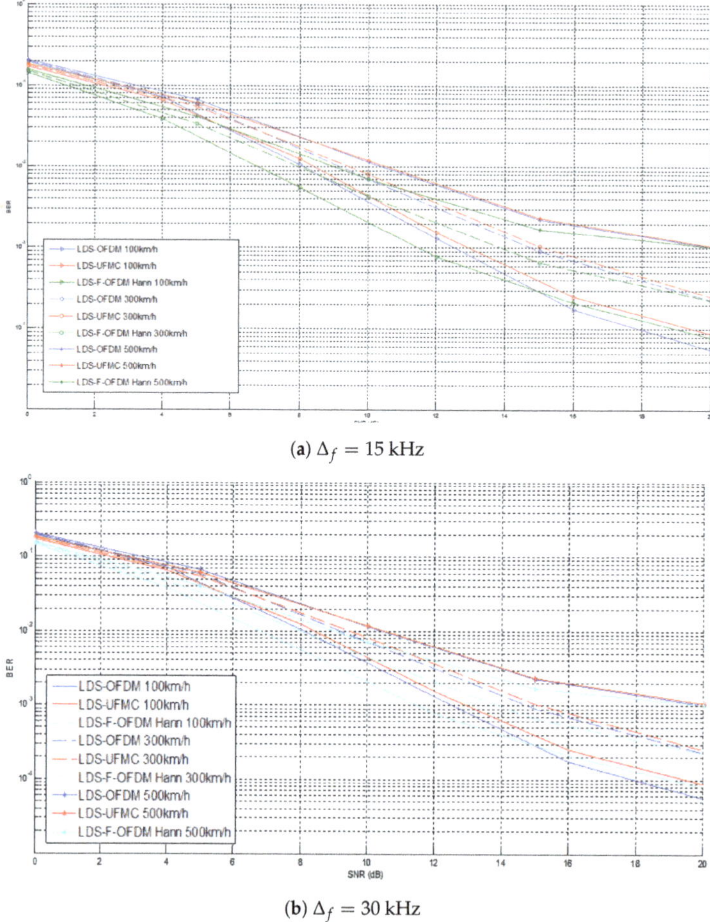

(**a**) $\Delta_f = 15$ kHz

(**b**) $\Delta_f = 30$ kHz

Figure 7. Schemes comparison over the EVA channel with different speed limits.

5. Conclusions

In this paper, we have proposed two efficient multiple access techniques namely LDS-F-OFDM and LDS-UFMC in which the LDS structure is combined with the new 5G waveforms. First of all, we have presented the state-of-art of the previous work done on LDS and we have detailed the transmitters and receivers of these new schemes. Then, we have highlighted our contribution that consists in evaluating these new schemes over different types of channels, specifically, a vehicular channel with a high mobility. Simulation results show that LDS-F-OFDM significantly achieves higher performance improvements compared to LDS-OFDM and LDS-UFMC in all scenarios, while maintaining an affordable complexity at the transmitter and the receiver side. The improvements are directly related to the advantages that f-OFDM waveform offers by addressing the adequate filter. In future work, we propose to analyse these schemes with different types of receivers in manner to reduce the complexity and to improve the performances. Some of the receivers that can be found in the literature are the SIC-MPA where we combine SIC and MPA to reduce the overall complexity and the JSG receiver which is a very high complex receiver but provides a lot of improvements to the system. Furthermore, we consider evaluating theses schemes over different vehicular channels such as a confined channel and with different channel specifications considering the other operating band of FR1 i.e., 25 GHz.

Author Contributions: Conceptualisation, I.K., R.E., F.E. and N.I.; methodology, I.K., R.E. and F.E.; software, I.K.; validation, R.E., F.E. and N.I.; formal analysis, R.E. and F.E.; investigation, I.K., R.E. and F.E.; resources, R.E. and F.E.; writing—original draft preparation, I.K.; writing—review and editing, I.K.; supervision, R.E., F.E. and N.I.; project administration, R.E. and F.E.; funding acquisition, R.E. and F.E. All authors have read and agreed to the published version of the manuscript.

Funding: The present research work has been supported by the European project SECREDAS funded ECSEL. The preliminary results were partly supported by ELSAT project. ELSAT 2020 project is co-financed by the European Union with the European Regional Development Fund, the French state and the Hauts de France Region Council.

Conflicts of Interest: The authors declare no conflict of interest.

Abbreviations

The following abbreviations are used in this manuscript:

OFDM	Orthogonal Frequency Division Multiplexing
FBMC	Filter-Bank Multi-Carrier
UFMC	Universal Filtered Multi-Carrier
BER	Bit Error Rate
LTE	Long Term Evolution
3GPP	3rd Generation Partnership Project
SC-FDMA	Single Carrier Frequency Division Multiple Access
ETSI	European Telecommunications Standards Institute
NR-V2X	New-Radio V2X
CP	Cyclic Prefix
UL	Uplink
DL	Downlink
QPSK	Quadrature Phase Shift Keying
QAM	Quadrature Amplitude Modulation
IDMA	Interleaver Division Multiple Access
PDMA	Pattern Division Multiple Access
SCMA	Sparse Code Multiple Access
CDMA	Code Division Multiple Access
JSG-IOTA	Joint Sparse Graph-Isotropic Orthogonal Transfer Algorithm
MPA	Message Passing Algorithm
FEC	Forward Error Correction
EPA	Expectation Propagation Algorithm
IFFT	Inverse Fast Fourier Transform
LS	Least Square
FFT	Fast Fourier Transform
ISI	Inter Symbol Interference
OOB	Out Of Band
PDP	Power Delay Profile
EVA	Extended Vehicular A

References

1. Sun, S.; Hu, J.; Peng, Y.; Pan, X.; Zhao, L.; Fang, J. Support for vehicle-to-everything services based on LTE. *IEEE Wirel. Commun.* **2016**, *23*, 4–8. [CrossRef]
2. Molina-Masegosa, R.; Gozalvez, J. LTE-V for Sidelink 5G V2X Vehicular Communications: A New 5G Technology for Short-Range Vehicle-to-Everything Communications. *IEEE Veh. Technol. Mag.* **2017**, *12*, 30–39. [CrossRef]

3. Gonzalez-Martín, M.; Sepulcre, M.; Molina-Masegosa, R.; Gozalvez, J. Analytical Models of the Performance of C-V2X Mode 4 Vehicular Communications. *IEEE Trans. Veh. Technol.* **2019**, *68*, 1155–1166. [CrossRef]
4. Festag, A. Standards for vehicular communication—From IEEE 802.11p to 5G. *Elektrotech. Inftech.* **2015**, *132*, 409–416. [CrossRef]
5. Anwar, W.; Franchi, N.; Fettweis, G. Physical Layer Evaluation of V2X Communications Technologies: 5G NR-V2X, LTE-V2X, IEEE 802.11bd, and IEEE 802.11p. In Proceedings of the 2019 IEEE 90th Vehicular Technology Conference (VTC2019-Fall), Honolulu, HI, USA, 22–25 September 2019; pp. 1–7.
6. 3GPP TR 37.985. Overall Description of Radio Access Network (RAN) Aspects for Vehicle-to-Everything (V2X) Based on LTE and NR. 3rd Generation Partnership Project; Technical Specification Group Radio Access Network. Available online: https://www.3gpp.org (accessed on 2 April 2020).
7. 3GPP TR 38.812. Study on Non-Orthogonal Multiple Access (NOMA) for NR. 3rd Generation Partnership Project; Technical Specification Group Radio Access Network. Available online: https://www.3gpp.org (accessed on 2 April 2020).
8. Ping, L.; Liu, L.; Wu, K.; Leung, W.K. Interleave division multiple-access. *IEEE Trans. Wirel. Commun.* **2006**, *5*, 938–947. [CrossRef]
9. Chen, Y.; Schaich, F.; Wild, T. Multiple Access and Waveforms for 5G: IDMA and Universal Filtered Multi-Carrier. In Proceedings of the 2014 IEEE 79th Vehicular Technology Conference (VTC Spring), Seoul, Korea, 18–21 May 2014; pp. 1–5.
10. Li, B.; Du, R.; Kang, W.; Liu, G. Multi-User Detection for Sporadic IDMA Transmission Based on Compressed Sensing. *Entropy* **2017**, *19*, 334.
11. Zeng, J.; Li, B.; Su, X.; Rong, L.; Xing, R. Pattern division multiple access (PDMA) for cellular future radio access. In Proceedings of the 2015 International Conference on Wireless Communications & Signal Processing (WCSP), Nanjing, China, 15–17 October 2015; pp. 1–5.
12. Ren, B.; Wang, Y.; Dai, X.; Niu, K.; Tang, W. Pattern matrix design of PDMA for 5G UL applications. *China Commun.* **2016**, *13*, 159–173. [CrossRef]
13. Nikopour, H.; Yi, E.; Bayesteh, A.; Au, K.; Hawryluck, M.; Baligh, H.; Ma, J. SCMA for downlink multiple access of 5G wireless networks. In Proceedings of the 2014 IEEE Global Communications Conference, Austin, TX, USA, 8–12 December 2014; pp. 3940–3945.
14. Liu, B.; Zhang, L.; Xin, X. Non-orthogonal optical multicarrier access based on filter bank and SCMA. *Opt. Express* **2015**, *23*, 27335–27342. [CrossRef] [PubMed]
15. Hoshyar, R.; Wathan, F.P.; Tafazolli, R. CTH06-4: Novel Low-Density Signature Structure for Synchronous DS-CDMA Systems. In Proceedings of the IEEE Globecom 2006, San Francisco, CA, USA, 28–30 November 2006; pp. 1–5.
16. Hoshyar, R.; Razavi, R.; Al-Imari, M. LDS-OFDM an Efficient Multiple Access Technique. In Proceedings of the 2010 IEEE 71st Vehicular Technology Conference, Taipei, Taiwan, 16–19 May 2010; pp. 1–5.
17. Wen, L.; Xiao, P.; Razavi, R.; Imran, M.A.; Al-Imari, M.; Maaref, A.; Lei, J. Joint Sparse Graph for FBMC/OQAM Systems. *IEEE Trans. Veh. Technol.* **2018**, *67*, 6098–6112. [CrossRef]
18. Vakilian, V.; Wild, T.; Schaich, F.; Brink, S.T.; Frigon, J.F. Universal filtered multi-carrier technique for wireless systems beyond LTE. In Proceedings of the IEEE Globecom Workshops (GC Wkshps), Atlanta, TX, USA, 9–13 December 2013.
19. Abdoli, J.; Jia, M.; Ma, J. Filtered OFDM: A new waveform for future wireless systems. In Proceedings of the 2015 IEEE 16th International Workshop on Signal Processing Advances in Wireless Communications (SPAWC), Stockholm, Sweden, 28 June–1 July 2015; pp. 66–70.
20. Zhang, L.; Ijaz, A.; Xiao, P.; Molu, M.M.; Tafazolli, R. Filtered OFDM Systems, Algorithms, and Performance Analysis for 5G and Beyond. *IEEE Trans. Commun.* **2018**, *66*, 1205–1218. [CrossRef]
21. Thakre, A. Optimal Filter Choice for Filtered OFDM. In Proceedings of the 2019 3rd International Conference on Electronics, Communication and Aerospace Technology (ICECA), Coimbatore, India, 12–14 June 2019; pp. 1035–1039. [CrossRef]
22. Hoshyar, R.; Wathan, F.P.; Tafazolli, R. Novel Low-Density Signature for Synchronous CDMA Systems Over AWGN Channel. *IEEE Trans. Signal Process.* **2008**, *56*, 1616–1626. [CrossRef]
23. 3GPP TR 38.886. V2X Services Based on NR; User Equipment (UE) Radio Transmission and Reception. 3rd Generation Partnership Project; Technical Specification Group Radio Access Network. Available online: https://www.3gpp.org (accessed on 2 April 2020). [CrossRef]

24. 3GPP TR 38.900. Study on Channel Model for Frequency Spectrum above 6 GHz. 3rd Generation Partnership Project; Technical Specification Group Radio Access Network. Available online: https://www.3gpp.org (accessed on 2 April 2020).
25. 3GPP TS 36.101. Evolved Universal Terrestrial Radio Access (E-UTRA); User Equipment (UE) Radio Transmission and Reception. 3rd Generation Partnership Project; Technical Specification Group Radio Access Network. Available online: https://www.3gpp.org (accessed on 2 April 2020).

Article

An Overview of FIR Filter Design in Future Multicarrier Communication Systems

Lei Jiang , Haijian Zhang *, Shuai Cheng, Hengwei Lv and Pandong Li

Signal Processing Laboratory, School of Electronic Information, Wuhan University, Wuhan 430072, China; teki97@whu.edu.cn

* Correspondence: haijian.zhang@whu.edu.cn

Received: 19 February 2020; Accepted: 27 March 2020; Published: 31 March 2020

Abstract: Future wireless communication systems are facing with many challenges due to their complexity and diversification. Orthogonal frequency division multiplexing (OFDM) in 4G cannot meet the requirements in future scenarios, thus alternative multicarrier modulation (MCM) candidates for future physical layer have been extensively studied in the academic field, for example, filter bank multicarrier (FBMC), generalized frequency division multiplexing (GFDM), universal filtered multicarrier (UFMC), filtered OFDM (F-OFDM), and so forth, wherein the prototype filter design is an essential component based on which the synthesis and analysis filters are derived. This paper presents a comprehensive survey on the recent advances of finite impulse response (FIR) filter design methods in MCM based communication systems. Firstly, the fundamental aspects are examined, including the introduction of existing waveform candidates and the principle of FIR filter design. Then the methods of FIR filter design are summarized in details and we focus on the following three categories—frequency sampling methods, windowing based methods and optimization based methods. Finally, the performances of various FIR design methods are evaluated and quantified by power spectral density (PSD) and bit error rate (BER), and different MCM schemes as well as their potential prototype filters are discussed.

Keywords: multicarrier modulation; prototype filter design; frequency sampling methods; windowing based methods; optimization based methods

---

1. Introduction

The rapid increase of mobile devices and the emergence of new technologies as well as services demand more efficient wireless cellular networks [1–4]. The 5th generation (5G) communication networks need to support abundant business scenarios, such as Internet of Vehicles [5–7], Internet of Things [8–10], virtual reality [11,12], device-to-device communications [13–16], and so forth. Therefore, plenty of important technologies are worth studying in future communication networks, for example, multicarrier modulation, massive multiple-input multiple-output (MIMO), millimeter wave, and so forth [1–4]. Due to the critical role of the prototype filter played in multicarrier modulations, that is, it determines the system performance such as stopband attenuation, inter-symbol interference (ISI), inter-channel interference (ICI) and phase noise caused by high operating frequencies, herein various prototype filter design methods in multicarrier communication networks are summarized.

It is known that cyclic-prefix orthogonal frequency division multiplexing (CP-OFDM) as the air interface of the 4th generation (4G) communication networks is capable of avoiding ICI and ISI. The modulated and demodulated signals are respectively generated through inverse fast Fourier transform (IFFT) and fast Fourier transform (FFT), thereby greatly reducing the system complexity and improving the signal transmission rate. OFDM has been used in conjunction with other technologies, for example, wavelet OFDM (WOFDM) [17,18], orthogonal frequency division multiple access (OFDMA) [19–21], Alamouti coded OFDM (AC-OFDM) [22–25], and MIMO-OFDM [26–28]. Although OFDM has been widely applied, it also has the following limitations: serious out-of-band leakage characteristic; strict synchronization and orthogonality among the subcarriers are needed; high peak to average power ratio (PAPR) resulting in non-linear distortion of the signal; sensitive to frequency offset, which has a significant impact on the system performance [29,30].

The above situations drive the urgency to conceive an appropriate modulation scheme in future communication networks. Faced with the requirements of the ultra-low latency, high spectrum efficiency, high transmission rate and business diversity in the future networks, researchers and practitioners in related fields have exerted a tremendous fascination on a number of alternative single-carrier or multicarrier modulation techniques [31], including CP-Discrete Fourier Transform spread OFDM (CP-DFT-s-OFDM) [32], filter bank multicarrier (FBMC) [33–39], generalized frequency division multiplexing (GFDM) [40,41], universal filtered multicarrier (UFMC) [42–44], and filtered-orthogonal frequency division multiplexing (F-OFDM) [45–47], and so forth. Apart from typical waveform design methods where waveform parameters are obtained manually, with the development of machine learning, waveform design methods with data-driven models have attracted increasing attention [48,49]. These modulation waveforms have their own merits and drawbacks in 5G communication scenarios, respectively. The prototype filter determines the performance of a specific modulation waveform, thus the evaluation standard of prototype filter is primarily characterized by minimizing the stopband energy, minimizing the maximum stopband ripple, minimizing the total interference (ISI and ICI) [50,51], and so forth. Digital filters are very important in digital communications and are generally divided into two categories—infinite impulse response (IIR) filter and finite impulse response (FIR) filter [52]. Based on the fact that only linear phase FIR filters are suitable for wireless communication systems [53], the FIR filter design has been the major research topic to realize prototype filters in the literature.

There have already been a few representative survey works related to prototype filters. In Reference [4], a survey of multicarrier communications about prototype filters, lattice structures and the implementation aspects is reported by A. Şahin et al., and this work provides four classes of filters according to the design criteria—energy concentration, rapid decay, spectrum nulling and channel/hardware characteristics. In Reference [54], the methods of prototype filter design for cosine modulated filter banks (CMFBs) with nearly perfect reconstruction (NPR) are reviewed by K. Shaeen et al., and these methods are categorized into nonlinear optimization methods [55–63], spectral factorization methods [64], linear search methods [65–70], interpolated finite impulse response (IFIR) methods [71–73], and frequency response masking (FRM) methods [58,74–79], and so forth. The authors of Reference [80] provide the basic introduction of the FIR filter, some relevant works on FIR and various factors which effect the performance of FIR filter in communication systems.

In this paper, we pay more attention to the literature from more recent years and provide a survey on various FIR filter design methods in multicarrier systems. Compared with the existing surveys in References [4,54,80], this paper provides an in-depth introduction to advanced filter design methods, which are not categorized based on the design criteria. In addition, this work can serve as an extension of the survey in Reference [54], which is dedicated to the aspect of CMFBs. Specifically, we firstly introduce the FIR filter design criteria, that is, the objective to be optimized. The evaluation criteria and the implementation methods are summarized in canonical form. Subsequently, the design methods of realizing the objectives are divided into three categories: frequency sampling methods, windowing based methods and optimization based methods. Lastly, the reviewed filter design methods are analyzed for multicarrier modulation schemes in terms of power spectral density (PSD) [81,82], bit error rate (BER) [83–85], spectral efficiency, latency, computational complexity, and so forth, which are important measurements to characterize the performance of modulation waveforms. We expect that this survey can provide a reference on how to select multicarrier modulation schemes and their corresponding prototype filters in future wireless communication systems.

The remainder of this overview paper is organized as follows. Section 2 introduces the concepts of different multicarrier modulations and the evaluation criteria of FIR filter design. In Section 3, the classification of FIR prototype filter design methods and corresponding implementation procedures are described in details. The discussions of different filter design methods in multicarrier modulation systems in terms of PSD and BER performances are presented in Section 4. Finally, Section 5 concludes the paper.

2. Multicarrier Modulations and FIR Filter

By splitting the transmitting data into several components and sending each of these components over separate carrier signals, multicarrier modulation (MCM) has numerous advantages compared to single carrier modulation, including the relative immunity to multipath fading, less susceptibility to interference caused by impulse noise, and enhanced immunity to ISI. In the following subsection, five important MCM waveforms—OFDM, FBMC, GFDM, UFMC and F-OFDM are introduced.

2.1. Modulation Waveforms

2.1.1. OFDM

Compared with conventional OFDM, windowed orthogonal frequency division multiplexing (W-OFDM) has a similar transceiver structure but non-rectangular transmit windows are utilized to smooth the edges of rectangular pulse, accordingly provide better spectral localization and reduce ICI [86]. For a raised cosine shape based non-rectangular window used in W-OFDM, the CP needs to be extended to maintain the orthogonality and the spectral efficiency will decrease for W-OFDM compared to OFDM [87]. In order to utilize non-contiguous spectrum fragments, resource block filtered OFDM (RB-F-OFDM) aims to split the available spectrum fragments into several resource blocks which make a chunk of some contiguous subcarriers. It generates and filters the signal transmitted on each resource block individually [88]. Analogous to the subcarrier filtering based modulation, spectrum leakage among these resource blocks is unavoidable in the case that the spacing between these blocks is narrow [89]. In CP-OFDM with weighted overlap and add (WOLA), some data parts are copied and added to the right and left part of conventional OFDM, then a pulse with soft edges takes place in the rectangular prototype filter, in the meanwhile, the soft edges are added to the cyclic extension by a time domain windowing and this results in a sharper side-lobe decay in frequency domain [90]. Therefore, CP-OFDM with WOLA is a special case of OFDM, aiming to improve the prototype filter such that it has more reasonable pulse shape used in regular CP-OFDM. Furthermore, for the sake of addressing the sacrifice of time resource resulting from added parts and avoiding possible data collision before transmission due to windowing process, overlap process is employed [91].

2.1.2. FBMC

The FBMC modulation technique, as one of the waveform candidates in 5G communication networks, has attracted a great amount of research attention. In the 1960s, the concept of FBMC modulation was firstly proposed by Chang and Saltherg [92]. However, it did not receive much attention by researchers due to its complexity. The well-known discrete multi-tone (DMT) modulation and discrete wavelet multi-tone (DWMT) modulation reported in the 1990s are two specific cases of FBMC modulation [93]. Currently, FBMC based systems have been extensively studied from the aspects of spectrum efficiency analysis [94–96], system complexity analysis [97], prototype filter design [98–101], frequency offset estimation [102,103], MIMO [104–106], and so forth. Ever-emerging research projects investigate the application of FBMC modulation in practical scenarios, for example, PHYDYAS [107], METIS [108], 5GNOW [109]. FBMC is robust against frequency offset because of its negligible spectrum leakage. Furthermore, FBMC does not need the guard band in frequency domain, which greatly improves the spectrum efficiency, thus FBMC can flexibly control the interference between adjacent subcarriers, and the synchronization requirement among subcarriers is relaxed [36].

Simultaneously, there are still some underlying issues of FBMC, for instance, high PAPR, which may well result in signal nonlinearities. Though many existing methods have been dedicated to lowing PAPR of FBMC [110–113], it is actually a trade-off between low PAPR and other characteristics, that is, in Reference [110], the PAPR reduction is at the expense of higher computation complexity. In spite of this, advantages above still make FBMC popular in the academic field, and FBMC has already been considered as the waveform candidate in future MCM networks [33].

2.1.3. GFDM

The GFDM modulation waveform, as a new 5G multicarrier modulation technique, was proposed by Fettweis et al. in 2009 [41]. Compared to OFDM, GFDM aiming to generalize traditional OFDM is based on separate block modulation, which makes it more flexible by configuring different subcarriers and symbols. To accommodate various types of services, GFDM can work with different prototype filters and different types of CPs. The design of the conforming prototype filter reduces the sidelobe interference, and the modulation process is converted from linear convolution to cyclic convolution through tail biting operation, thus shortening the length of CP. Similar to FBMC, GFDM is also robust against time-frequency offset but has a low PAPR. However, each subcarrier in GFDM modulation does not keep the orthogonality in the frequency domain and ICI is introduced even under ideal channel, which increases the complexity of the receiver algorithm and meanwhile raises BER.

2.1.4. UFMC

UFMC modulation technique was proposed by Vida Vakilian et al. to solve the ICI problem in OFDM systems [42]. UFMC modulation enjoys the following advantages of relaxing the requirement of CP, having high spectrum efficiency, relaxing carrier synchronization, and being suitable for fragmented debris spectrum utilization. In addition, UFMC supports short burst asynchronous communication because the filter lengths depend on the sub-band widths. Nevertheless, similar to OFDM, the performance of UFMC is also inevitably affected by the carrier frequency offset (CFO), which has attracted a lot of research interest in order to mitigate the significant impact of CFO on UFMC based system performance [114–116].

2.1.5. F-OFDM

The basic principle of F-OFDM technique is to divide the carrier bandwidth of OFDM into sub-bands with different parameters, filter the sub-bands, and leave less isolation bands in the sub-bands [46]. Firstly, to support diverse businesses, F-OFDM modulation supports flexible sub-band configurations for different subcarrier spacing. Secondly, it supports different sub-band configurations allowing different CP lengths to better adapt transmission channels, and different sub-bands on asynchronous signal transmission are also supported, thus saving the signaling overhead. Finally, with better out of band suppression characteristics compared to OFDM, it saves the cost of protection zone, that is, the spectral efficiency is improved. However, filters need to be dynamically designed for each fragment, which makes the use of F-OFDM systems challenging [88].

2.2. *FIR Filter*

In this subsection, we introduce the basic principle of digital FIR filters as well as the evaluation criteria of a designed filter. It is known that the condition of distortionless transmission and filtering is that the amplitude response of the system should be constant in the effective spectral range of the signal, and the phase response should be a linear function of the frequency (i.e., linear phase). The prototype filter is required to have linear phase characteristic in wireless communication applications. In contrast to the IIR filter, the FIR filter has the ability to achieve linear phase filtering [117]. In addition, as an all-zero filter, the hardware and software structures of the FIR filter could be established without considering the stability problem. Therefore, FIR filters have been widely used in the field of wireless communications [53].

Assuming the length of the impulse response $h(n)$ of a linear phase FIR filter is N, then the frequency response function is defined as

$$H(e^{j\omega}) = \sum_{n=0}^{N-1} h(n)e^{-j\omega n}, \tag{1}$$

where ω is the angular frequency. According to the parity of N and the symmetry of $h(n)$, there are four types of FIR transfer functions. The condition for the most widely used low-pass FIR filter is given by

$$h(n) = h(N-1-n), \quad 0 \leq n \leq N-1. \tag{2}$$

The ripples are observed in both the passband and stopband, and the maximum approximation error δ_p is described by

$$1-\delta_p \leq\mid H(e^{j\omega}) \mid\leq 1, \quad \mid \omega \mid\leq \omega_p. \tag{3}$$

The amplitude of the stopband is approximated by the maximum error δ_s, that is,

$$\mid H(e^{j\omega}) \mid\leq \delta_s, \quad \omega_s \leq\mid \omega \mid\leq \pi, \tag{4}$$

where ω_p and ω_s are passband and stopband boundary frequencies, δ_p and δ_s are passband and stopband ripples, respectively (Gibbs effect), and $\omega_s - \omega_p$ is the transition bandwidth. In order to make designed filters close to the ideal low-pass filter, several evaluation criteria are given in [50,51], including least-squares (LS), minimax, peak-constrained least-squares (PCLS), minimum total interference (ICI and ISI), and and so forth.

- **LS criterion**

  The goal of the least-squares criterion is to minimize the stopband energy of the filter, whose objective function is

$$J = \int_{\omega_s}^{\pi} \mid H(e^{j\omega}) \mid^2 d\omega. \tag{5}$$

- **Minimax criterion**

  The goal of the minimax criterion is to minimize the maximum stopband ripple, and its objective function can be written as

$$J = \max_{\omega\in[\omega_s,\pi]} \mid H(e^{j\omega}) \mid. \tag{6}$$

- **PCLS criterion**

  The PCLS criterion establishes a trade-off between the LS and the minimax criteria. The PCLS criterion can be described as below

$$\begin{aligned} J &= \textstyle\int_{\omega_s}^{\pi} \mid H(e^{j\omega}) \mid^2 d\omega \\ \text{s.t.} &\mid H(je^{j\omega}) \mid\leq \delta, \end{aligned} \tag{7}$$

  where δ is a prescribed value. If δ is close to zero, the PCLS criterion approaches the minimax criterion, while in the limit of δ, that is, δ is up to infinite, the criterion turns out the LS criterion.

- **Minimum total interference criterion**

  This criterion is to minimize the total interference of ICI and ISI for filter bank structure. Its objective function is defined as

$$J = \text{ISI} + \text{ICI}, \tag{8}$$

  where

$$\text{ISI} = \max_k \left(\sum_n ([T_{\text{TMUX}}(n)]_{k,k} - \delta(n-\triangle))^2 \right), \tag{9}$$

$$\text{ICI} = \max_k \left(\sum_{l=0, l \neq k}^{N-1} \sum_n ([T_{\text{TMUX}}(n)]_{k,l})^2 \right), \tag{10}$$

where $T_{\text{TMUX}}(n)$ is the transfer matrix, the element $[T_{\text{TMUX}}(n)]_{a,b}$ represents the relationship between the input signal $X(z^N)$ and the output signal $Y(z^N)$, where z is the complex argument of z-transform. $\delta(n)$ is the ideal impulse, and $\triangle$ denotes the delay of the TMUX system. Thus the transfer function between input and output signals is given by

$$Y(z^N) = T_{\text{TMUX}}(z^N) \cdot X(z^N), \tag{11}$$

where

$$X(z^N) = [X_0(z) \quad X_1(z) \quad \cdots \quad X_{N-1}(z)]^T, \tag{12}$$

$$Y(z^N) = [Y_0(z) \quad Y_1(z) \quad \cdots \quad Y_{N-1}(z)]^T. \tag{13}$$

3. FIR Prototype Filter Design

As discussed in Section 2, there are some design criteria to customize the performance of FIR. Thus, there must be some specific methods of FIR design corresponding to these criteria. In this work, the filter design methods are divided into three major categories: the frequency sampling methods, the windowing based methods, and the optimization based methods. For each type of method, we firstly introduce their basic concepts and subsequently some typical design examples are given. At last the advantages and disadvantages of the summarized methods are discussed.

3.1. Frequency Sampling Methods

The idea of frequency sampling is conceived from the frequency domain perspective. This kind of method takes uniform spacing sampling of an ideal frequency response $H_d(e^{j\omega})$, which is given by

$$H_d(k) = H_d(e^{j\omega})|_{\omega = \frac{2\pi}{N}k}. \tag{14}$$

Then $H_d(k)$ is used as the sampled values of the actual linear phase FIR filter, written as

$$H(k) = H_d(k), \quad k = 0, 1, \cdots, N-1. \tag{15}$$

The N-point inverse discrete Fourier transform (IDFT) for $H(k)$ yields the following impulse response

$$\begin{aligned} h(n) =& \text{IDFT}\Big[H(k)\Big] \\ =& \frac{1}{N} \sum_{k=0}^{N-1} H(k) W_N^{-kn}, \quad n = 0, 1, \cdots, N-1, \end{aligned} \tag{16}$$

where $W_N = e^{-j\frac{2\pi}{N}}$.

The exponential form of $H(k)$ is given by

$$H(k) = A(k) e^{j\theta(k)}, \tag{17}$$

where $A(k)$ and $\theta(k)$ are amplitude sampling and phase sampling, respectively. To design a linear phase FIR filter in practical applications, the condition in Equation (2) should be satisfied. Thus certain constraint conditions are imposed on $A(k)$

$$\begin{cases} A(k) = A(N-k), & N \text{ is odd}, \\ A(k) = -A(N-k), & N \text{ is even}, \end{cases} \tag{18}$$

and also on $\theta(k)$

$$\theta(k) = -\omega\left(\frac{N-1}{2}\right)\Bigg|_{\omega=\frac{2\pi k}{N}} = -\left(\frac{N-1}{N}\right)\pi k. \tag{19}$$

A series of frequency sampling based filter design methods have been reported according to the above formulations. Some of commonly used methods are described in the following subsections.

3.1.1. Bellanger's Method

In [118], Bellanger et al. use a typical frequency sampling method to design a filter, which is chosen as the prototype filter of FBMC modulation system in PHYDYAS project. Specific implementation is as follows [119]: assuming an integer K, the number of subcarriers is F and the number of samples is KF, two conditions are proposed:

Condition 1: To approximately meet the Nyquist criterion, the following equation should be satisfied

$$\begin{cases} H_0 = 1, & \\ H_k^2 + H_{K-k}^2 = 1, & \\ H_{KF-k} = H_k, & 1 \le k \le K-1, \\ H_k = 0, & K \le k \le KF-K, \end{cases} \tag{20}$$

where $H_k(0 \le k \le KF-1)$ is the kth frequency weighting coefficient.

Condition 2: To ensure stopband performance, the following equation is satisfied

$$H_0 + 2\sum_{k=1}^{K-1}(-1)^k H_k = 0. \tag{21}$$

When $K = 3$ and 4, the filter weighting coefficients are

$$\begin{cases} K=3: H_0 = 1; H_1 = 0.9144; H_2 = 0.4114, \\ K=4: H_0 = 1; H_1 = 0.9720; H_2 = \sqrt{2}; H_3 = 0.2351. \end{cases} \tag{22}$$

Then the impulse response is obtained by IDFT

$$h(t) = \begin{cases} 1 + 2\sum_{k=1}^{K-1}(-1)^k H_k \cos(\frac{2\pi t}{KT}), & -\frac{KT}{2} \le t \le \frac{KT}{2}, \\ 0, & \text{otherwise}. \end{cases} \tag{23}$$

3.1.2. Viholainen's Method

Similar to the design in [118], **Condition 1** is also considered in Viholainen's method [51], but **Condition 2** is relaxed and it is assumed that $H_1 = \chi$. When $K = 3$ and $K = 4$, the filter weighting coefficients are

$$\begin{cases} K=3:\ H_0 = 1;\ H_1 = \chi;\ H_2 = \sqrt{1-\chi^2}, \\ K=4:\ H_0 = 1;\ H_1 = \chi;\ H_2 = 1/\sqrt{2};\ H_3 = \sqrt{1-\chi^2}. \end{cases} \tag{24}$$

According to a specific objective function, the optimal solution can be obtained by a simple global search algorithm.

In [51], the LS criterion and the minimax criterion are selected for comparison with the method in [118]. The frequency responses of prototype filters in [118] and [51] based on the LS criterion are shown in Figure 1a, where the number of subcarriers is 16. As can be seen, the side-lobes of the prototype filters when $K = 4$ are lower than those with $K = 3$, that is, better stopband performance is benefited by a larger value of K. Furthermore, the first side-lobe of the prototype filter in [51] is a little lower than that in [118]. Figure 1b shows the frequency responses of prototype filters in [118] and [51] based on the minimax criterion. Similarly, the prototype filters in [118] have higher first side-lobe levels. It is also observed from Figure 1, the prototype filters based on the minimax criterion improve the stopband performance at the cost of sacrificing the main-lobe width.

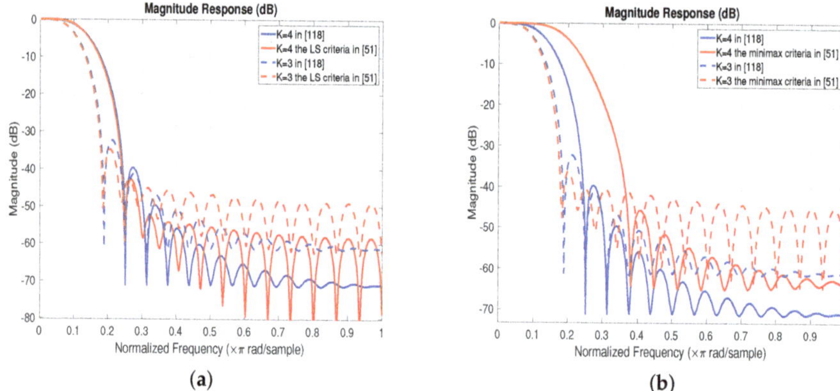

Figure 1. (**a**) Frequency responses of prototype filters implemented based on the least squares (LS) criterion in [51] and the method in [118] (K = 3 and K = 4). (**b**) Frequency responses of prototype filters based on the minimax criterion in [51] and the method in [118] (K = 3 and K = 4).

3.1.3. Cruz-Roldán's Method I

A frequency sampling method for arbitrary length FIR filters is proposed in [120]. Define the frequency response of the prototype filter $H(e^{j\omega})$, let $H[k] = H(e^{j\omega_k})$, where $\omega_k = (k+\alpha)\cdot 2\pi/N$, $0 \le k \le (N-1)$ and $\alpha = 0$ or $1/2$. The authors in [120] assume $h(n)$ is real and symmetric, thus the filter coefficients are obtained according to [117] when $\alpha = 0$

$$h[n] = \frac{1}{N}\left\{P[0] + 2\sum_{k=1}^{\lfloor N/2\rfloor -1} P[k]\cos\left((n+1/2)\cdot\frac{2\pi k}{N}\right)\right\}, \tag{25}$$

where $P[k] = H[k]\cdot e^{-jk\pi/N}$, $\lfloor \mathrm{M} \rfloor$ denotes the largest integer less than M. While if $\alpha = 1/2$, the filter is expressed as

$$h[n] = \frac{2}{N}\sum_{k=0}^{\lfloor N/2\rfloor -1} P[k]\sin\left((n+1/2)\cdot\frac{2\pi}{N}\cdot(k+1/2)\right), \tag{26}$$

where $P[k] = H[k]\cdot e^{-j(N-2k-1)\pi/(2N)}$.

Initial values of $H[k]$ are determined before optimization, and the center of the transition band is $\omega_r = (r+\alpha)\cdot 2\pi/N, r \in Z^+$. In the transition band, the number of samples L $(L>1)$ should be defined. Thus the magnitude response is defined as

$$\mid H[k] \mid = \begin{cases} 1, & 0 \le k \le r - \lceil L/2 \rceil, \\ & \text{(passband)} \\ f(k), & r - \lceil L/2 \rceil + 1 \le k \le r + \lfloor L/2 \rfloor, \\ & \text{(transition band)} \\ 0, & r + \lfloor L/2 \rfloor + 1 \le k \le \lfloor (N-1)/2 \rfloor, \\ & \text{(stopband)} \end{cases} \tag{27a}$$

$$\mid H[N-1-k] \mid = \mid H[k] \mid, \quad \lfloor (N-1)/2 \rfloor + 1 \le k \le N-1, \tag{27b}$$

where $\lceil M \rceil$ denotes the smallest integer more than M. And the phase response $\arg\{H[k]\}$ is defined as

$$\arg\{H[k]\} = \begin{cases} -\dfrac{N-1}{2}\cdot\dfrac{2\pi}{N}\cdot(k+\alpha), \\ \qquad 0 \le k \le \lfloor (N-1)/2 \rfloor, \\ \dfrac{N-1}{2}\cdot\dfrac{2\pi}{N}\cdot\Big(N-(k+\alpha)\Big), \\ \qquad \lfloor (N-1)/2 \rfloor + 1 \le k \le N-1. \end{cases} \tag{28}$$

The function $f(k)$ in (27a) is to obtain the magnitude values of the transition band samples. In [121], $f(k)$ is chosen as

$$f(k) = \frac{\omega_s - k\cdot 2\pi/N}{\omega_s - \omega_p}. \tag{29}$$

While in [120] $f(k)$ has a different expression as below

$$f(k) = 0.95 - \left(\frac{\omega_s - (L+1-k)\cdot 2\pi/N}{\omega_s - \omega_p}\right)^2. \tag{30}$$

When the initial values are determined, the specific steps of the optimization procedure are as follows:

(a) Initialize the filter length N and the required number of samples L in the transition band.

(b) Initialize the frequency response in (27) and (28). The resulting vector $|\mathbf{H}[k]|$ is presented as follows

$$|\mathbf{H}_{opt}[k]| = \left[\underbrace{1 \ldots 1}_{\text{passband}} \;\; \underbrace{f[q+1] \ldots f[q+L]}_{\text{transition}} \;\; \underbrace{0 \ldots 0}_{\text{stopband}} \right], \tag{31}$$

where $0 \le k \le \lfloor (N-1)/2 \rfloor$, $q = r - \lceil L/2 \rceil$.

(c) Let $\mathbf{f} = \Big[f[q+1] \; f[q+2] \; \cdots f[q+L] \Big]$ be the vector whose elements are the samples of the magnitude response at the transition band. Find $\mathbf{f}_{opt}$ and minimize an objective function ψ defined as [122]

$$\psi = \max_{n, n \neq 0} \mid g[2Mn] \mid, \tag{32}$$

where M is related to the number of channels, $G(e^{j\omega})$ is the DFT of $g[n]$ and defined as $G(e^{j\omega}) = \mid H(e^{j\omega}) \mid^2$.

(d) Calculate the optimum values of the frequency response samples $\mathbf{H}_{opt}[k]$. These values are obtained from (27) and (28), by replacing the initial values of the magnitude response in the transition band in (27a) by the optimised values $\mathbf{f}_{opt}$ obtained in the previous step

$$|\mathbf{H}_{opt}[k]| = \left[\underbrace{1 \ldots 1}_{\text{passband}} \quad \underbrace{f_{opt}[q+1] \ldots f_{opt}[q+L]}_{\text{transition}} \quad \underbrace{0 \ldots 0}_{\text{stopband}} \right], \tag{33}$$

$$|\mathbf{H}_{opt}[N-1-k]| = |\mathbf{H}_{opt}[k]|, \quad \lfloor (N-1)/2 \rfloor + 1 \leq k \leq N-1. \tag{34}$$

(e) Based on $\mathbf{H}_{opt}[k]$, the prototype filter coefficients are obtained through (25) or (26).

In [65], the objective function ψ in **step 3** is defined as

$$\psi = \max_{\omega} \left\{ |H(e^{j\omega})|^2 + |H(e^{j(\omega-\pi/M)})|^2 - 1 \right\}. \tag{35}$$

The minimization algorithms in MATLAB Optimization Toolbox [120] can be used to solve this optimization problem.

3.1.4. Cruz-Roldán's Method II

In [123], a multi-objective optimization technique based on the Cruz-Roldán's method I in [120] is proposed. The difference between [123] and [120] lies in the objective function, and the objective function of Cruz-Roldán's method II is

$$\begin{aligned} \psi &= \frac{1}{2\pi} \int_{\omega_s}^{2\pi-\omega_s} |H(e^{j\omega})|^2 d\omega \\ &= \mathbf{h}\mathbf{S}\mathbf{h}^{\mathrm{T}} \\ &= \frac{1}{N^2} \mathbf{H}^{\mathrm{T}} (\mathbf{W}_N^{-1})^{\mathrm{T}} \mathbf{S} \mathbf{W}_N^{-1} \mathbf{H}, \\ \text{s.t.} & \begin{cases} \max\limits_{\omega \in [0,\pi]} (|T_0(e^{j\omega})|) - \min\limits_{\omega \in [0,\pi]} (|T_0(e^{j\omega})|) \leq \delta_{pp}^{ini}, \\ MSA \geq MSA^{ini}, \end{cases} \end{aligned} \tag{36}$$

where $\mathbf{h} = [h(0), h(1), \cdots h(N-1)]$, $\mathbf{h}^{\mathrm{T}} = (1/N)\mathbf{W}_N^{-1}\mathbf{H}$, $\mathbf{W}_N^{-1}$ is the IDFT matrix, and the elements of $N \times N$ matrix $\mathbf{S}$ are given by

$$[\mathbf{S}_{i,j}] = \begin{cases} 1 - \dfrac{\omega_s}{\pi}, & i = j, \\ -\dfrac{\sin[\omega_s(i-j)]}{\pi(i-j)}, & \text{otherwise}. \end{cases} \tag{37}$$

The values of δ_{pp}^{ini} (amplitude distortion [121]) and MSA^{ini} (minimum stopband attenuation) are the initially fixed goals of the problem. The overall distortion transfer function $T_0(e^{j\omega})$ is obtained as

$$T_0(e^{j\omega}) = \frac{e^{-j\omega(N-1)}}{F} \sum_{k=0}^{2F-1} |H(e^{j(\omega - k \cdot \pi/F)})|^2, \tag{38}$$

where F is the number of subcarriers. This method requires the values of δ_{pp}^{ini} and MSA^{ini} to be selected appropriately, and the MATLAB function *fgoalattain* can solve the multi-objective optimization problem.

3.1.5. Salcedo-Sanz's Method

In [124], the authors propose the variable limits evolutionary programming (VLEP) based on the evolutionary programming (EP) to design the prototype filter, including variable limits classical evolutionary programming (VLCEP) and variable limits fast evolutionary programming (VLFEP). The final optimum solutions are sensitive to the initial values of $f(k)$ in [120], while the algorithm in [124] can solve this problem.

Three design methods in [120,123,124] are analyzed in terms of amplitude distortion peak, aliasing error and minimum stopband attenuation, as presented in Table 1. The number of subcarriers is 128 and the length of the prototype filter is 2049. We select the VLFEP algorithm in [124]. Generally, all of the three methods can obtain nearly perfect reconstruction (NPR) filter banks. By comparison, the method in [123] achieves the best filter design performance. To summarize, the essence, pros and cons of aforementioned frequency sampling methods are presented in Table 2.

Table 1. Analysis of prototype filter design in [120,123,124] by amplitude distortion peak, aliasing error, minimum stopband attenuation.

Prototype Filter Design	Amplitude Distortion Peak	Aliasing Error	Minimum Stopband Attenuation
Ref. [120]	7.0449×10^{-4}	−139.48 dB	78 dB
Ref. [123]	3.5001×10^{-5}	−151.39 dB	108 dB
Ref. [124]	1.2848×10^{-4}	−93.27 dB	69 dB

Table 2. Summary of frequency sampling methods.

Methods	Comments	Pros and Cons
Bellanger's method [118]	Design the filter parameter under the constraints of controlling stopband performance and satisfying Nyquist criterion.	**Pros:** Favorable stopband attenuation. **Cons:** Long filter length (Long latency).
Viholainen's method [51]	Optimize the filter parameters according to different evaluation criteria under the condition of Nyquist criterion.	**Pros:** Good stopband attenuation; Flexibly select suitable filter parameters according to different evaluation criteria. **Cons:** Need long filter length.
Cruz-Roldán's method I [120]	An optimization scheme based on frequency sampling is proposed to obtain the filter parameters.	**Pros:** Low computational complexity; Design a filter with arbitrary length. **Cons:** Difficult to choose the sampling values of transition band for fast convergence.
Cruz-Roldán's method II [123]	Based on the optimization algorithm proposed in [120], the objective function is improved to achieve multi-objective optimization.	**Pros:** The stopband energy and stopband attenuation are minimized simultaneously. **Cons:** Need to appropriately select initial parameters to ensure the performance.
Salcedo-Sanz's method [124]	The VLEP algorithm [124] is proposed based on the classical and fast evolutionary programming algorithm.	**Pros:** Robust to the initial conditions; The minimum of the objective function can be reliably obtained. **Cons:** High computational complexity.

3.2. *Windowing Based Methods*

The idea of windowing based methods is to use the digital FIR filter to approximate the desired filtering characteristics. Assuming that the desired filter frequency response function is $H_d(e^{j\omega})$, the impulse response is denoted as $h_d(n)$. Considering the windowing based methods focus on the perspective of time domain, the ideal $h_d(n)$ with a certain shape of window function is intercepted as $h(n)$ with finite length, whose frequency response $H(e^{j\omega})$ approximates the desired frequency response $H_d(e^{j\omega})$. The specific design steps of windowing based methods are as follows:

step 1: Taking linear phase low-pass FIR filter as an example, the general selection of $H_d(e^{j\omega})$ is

$$H_d(e^{j\omega}) = \begin{cases} e^{-j\omega\tau}, & |\omega| \leq \omega_c, \\ 0, & \omega_c \leq |\omega| \leq \pi, \end{cases} \tag{39}$$

where τ is a constant.

step 2: Determine $h_d(n)$ via IDFT

$$h_d(n) = \frac{1}{2\pi} \int_{-\omega_c}^{\omega_c} H_d(e^{j\omega}) e^{j\omega n} d\omega = \frac{\sin[\omega_c(n-\tau)]}{\pi(n-\tau)}. \tag{40}$$

step 3: The impulse response $h(n)$ of the linear phase FIR filter is obtained by multiplying a specific window $w(n)$, as below

$$h(n) = h_d(n)w(n). \tag{41}$$

Thus the corresponding $H(e^{j\omega})$ is obtained by DFT.

3.2.1. Jain's Method

Like the concept of Bartlett-Hanning window, a new window consisting of a Hamming window and a Blackman is proposed in [125], which is given by

$$\begin{aligned} w(n) =& \lambda \left(0.54 - 0.46 \cos \frac{2\pi n}{N-1} \right) \\ &+ (1-\lambda) \left(0.42 - 0.5 \cos \frac{2\pi n}{N-1} + 0.8 \cos \frac{4\pi n}{N-1} \right), \end{aligned} \tag{42}$$

where $0 \leq |n| \leq (N-1)/2$.

Note that the formula in (42) becomes a Hamming window if $\lambda = 1$ and a Blackman window if $\lambda = 0$. Figure 2 shows the frequency responses of Hamming window, Blackman window and this new window, where the variable λ is evaluated as 0.0625 to ensure superior performance than using Blackman window or Hamming window alone, and the length of the filters is $N = 128$. Compared with Blackman and Hamming windows, the new window has the best first side-lobe level, the best maximum side-lobe level and the best spectral efficiency.

3.2.2. Kumar's Method

In [126], the Hamming and Gaussian windows are combined as a new window function, whose expression is

$$w(n) = \left[0.54 + 0.46 \cos(\frac{2\pi n}{N-1}) \right] e^{-\frac{1}{2}(\frac{2\alpha n}{N-1})^2}, \tag{43}$$

where $0 \leq |n| \leq (N-1)/2$, and the value of the parameter α determines the performance of the filter.

Figure 3 shows the frequency responses of Hamming window, Gaussian window and the new window proposed in [126] for $N = 64$. Note that the window in [126] has the best side-lobe attenuation compare with Hamming and Gaussian windows. However, it increases the width of the main-lobe.

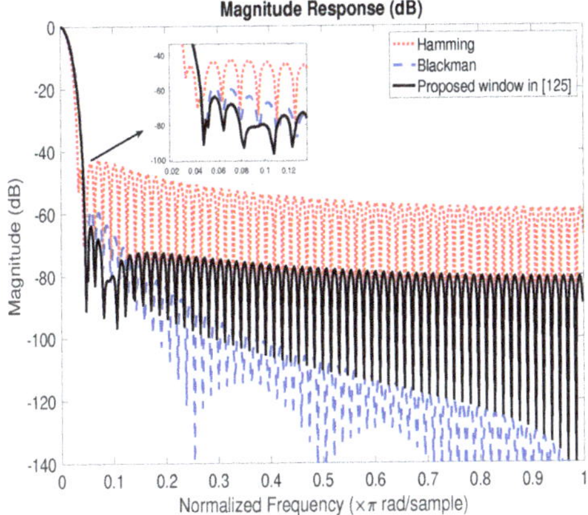

Figure 2. Frequency responses of Hamming window, Blackman window and the new window proposed in [125].

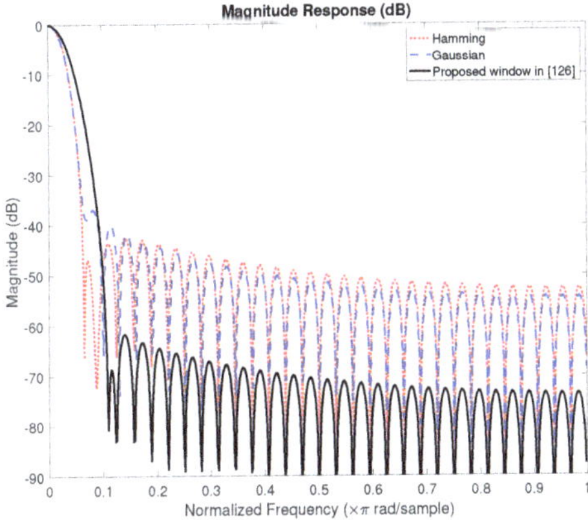

Figure 3. Frequency responses of Hamming window, Gaussian window and the new window proposed in [126].

3.2.3. Mottaghi-Kashtiban's Method

In [127], a special case of raised cosine windows is presented. The proposed window function is

$$w[n] = a_0 - a_1 \cos(\frac{2\pi n}{N-1}) - a_3 \cos(\frac{6\pi n}{N-1}),\ 0 \le n \le N-1. \tag{44}$$

For normalization, that is, $w\left[\frac{N-1}{2}\right] = 1$, then

$$a_0 + a_1 + a_3 = 1. \tag{45}$$

This new window is symmetric about $(N-1)/2$, and it has linear phase, thus the window function is obtained by

$$w[n] + w[n-(N-1)/2] = 2a_0,\ \frac{N-1}{2} \le n \le N-1. \tag{46}$$

The expression in Equation (44) is a 4th order raised cosine window [128] and its third term is zero

$$w[n] = \sum_{i=0}^{3} a_i \cos\left(\frac{2i\pi n}{N-1}\right), \quad 0 \le n \le M, \; a_2 = 0. \tag{47}$$

The optimal values are found by optimization and approximation. The relationship between the parameters in (47) and the length of window function N can be expressed as

$$\begin{cases} a_0 = 0.537 - \dfrac{0.3}{N+14}, \\ a_1 = 0.46 + \dfrac{0.25}{N+14}, \\ a_3 = 1 - a_0 - a_1. \end{cases} \tag{48}$$

Figure 4 shows the frequency responses of Hamming window, Gaussian window and the new window proposed in [127] for a typical filter length of $N = 41$. All of the windows have approximately equal mainlobe width. Compared with Hamming window, the new window has better peak level of maximum side-lobes, meanwhile it has better performance of side-lobe attenuation than Gaussian window.

3.2.4. Rakshit's Method

In [129], a new form of adjustable window function combining a tangent hyperbolic function and a weighted cosine series is proposed. In this paper, the modified tangent hyperbolic function is given as

$$\begin{aligned} y_1 = \tan &\left\{ \frac{n - \frac{N-1}{2} + \cosh^2(\alpha)}{B} \right\} \\ &- \tan \left\{ \frac{n - \frac{N-1}{2} - \cosh^2(\alpha)}{B} \right\}, \end{aligned} \tag{49}$$

and the weighted cosine function is expressed as

$$y_2 = 0.375 - 0.5\cos\left(\frac{2\pi n}{N-1}\right) + 0.125\cos\left(\frac{4\pi n}{N-1}\right), \tag{50}$$

where α and B are the constants, and the symbol $n = 0, 1, 2, 3, \cdots, (N-1)$. Then the new window function can be expressed as

$$w(n) = \begin{cases} (y_1 \times y_2)^{\gamma^{\frac{1}{5}}}, & 0 \le n \le N, \\ 0, & \text{otherwise}, \end{cases} \tag{51}$$

where γ is a variable which controls the shapes and frequency response of window function, and " $\times$ " denotes dot product.

Figure 5 shows the frequency responses of Gaussian window and the new window proposed in [129]. The length of the filters is $N = 65$, and $B = 1$, $\alpha = 2.5$, $\gamma = 1.5$ for the new window in [129]. It can be observed that the new window has much less side-lobe peak level than Gaussian window while its main-lobe width keeps exactly the same.

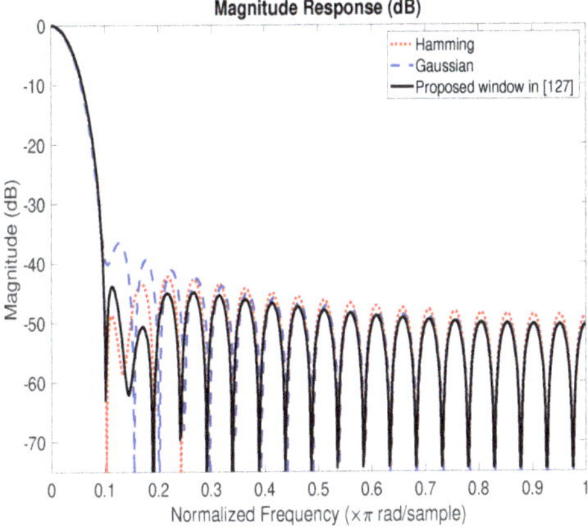

Figure 4. Frequency responses of Hamming window, Gaussian window and the new window proposed in [127].

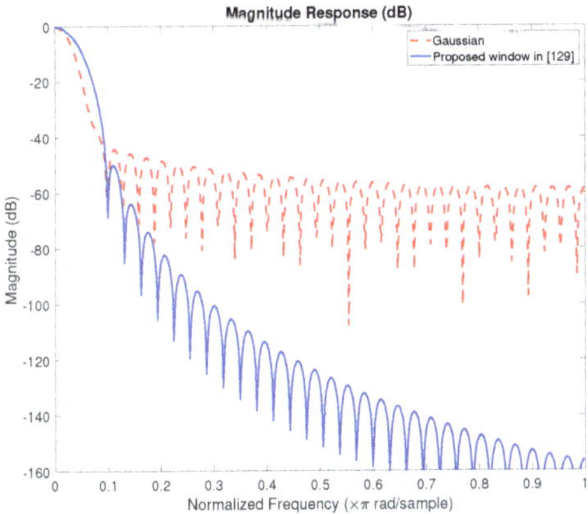

Figure 5. Frequency responses of Gaussian window and the new window proposed in [129].

3.2.5. Martin-Martin's Method

In [130], a window designing method for cosine-modulated transmultiplexers is proposed, which is called generalized windowing method for transmultiplexers (GWMT). A 4th order generalized cosine window function is expressed as

$$w(n) = \sum_{i=0}^{3}(-1)^i A_i \cos\left(\frac{2\pi in}{N-1}\right), \tag{52}$$

where $n = 0, 1, 2, \cdots, N-1$, and A_i are the weights of the terms for $i = 0, 1, 2, 3$. The designed window function is normalized as

$$\sum_{i=0}^{3} A_i = 1. \tag{53}$$

The weights A_i and the cut-off frequency ω_c of the ideal low-pass filter can be adjusted by minimizing the objective function

$$\phi(\mathbf{x}) = -\frac{1}{F}\sum_{f=0}^{F-1}\frac{E_f(\mathbf{x})}{\beta E_{\mathrm{ICI}}^{(f)}(\mathbf{x}) + (1-\beta)E_{\mathrm{ISI}}^{(f)}(\mathbf{x})}, \tag{54}$$

and

$$\begin{cases} E_f = \dfrac{1}{\pi}\displaystyle\int_0^{\pi}(|T_{ff}(e^{j\omega})|^2)d\omega, \\ E_{ICI}^{(f)} = \dfrac{1}{\pi}\displaystyle\int_0^{\pi}\left(\sum_{l=0,l\neq f}^{F-1}|T_{fl}(e^{j\omega})|^2\right)d\omega, \\ E_{ISI}^{(f)} = \dfrac{1}{\pi}\displaystyle\int_0^{\pi}\left\{\|T_{ff}\|_1 - |T_{ff}(e^{j\omega})|\right\}^2 d\omega, \\ \|T_{ff}\|_1 = \dfrac{1}{\pi}\displaystyle\int_0^{\pi}|T_{ff}(e^{j\omega})|d\omega, \end{cases} \tag{55}$$

where F is the channel number, β is the compromise factor satisfying $0 \leq \beta \leq 1$, T describes the transfer function, $\mathbf{x} = [A_0, A_1, A_2, \omega_c]$ denotes the adjustable parameter vector, and A_3 can be obtained by the Equation (53). This optimization problem is solved by the Nelder-Mead simplex minimization algorithm.

Figure 6 displays the frequency responses of Blackman window, Kaiser window and the generated window by GWMT method. The length of the filters is $N = 2KF$, where $K = 3$, $F = 32$. It is seen that the GWMT method based window has the highest signal-to-overall-interference ratio levels. The comments, pros and cons of the above windowing based methods are summarized in Table 3.

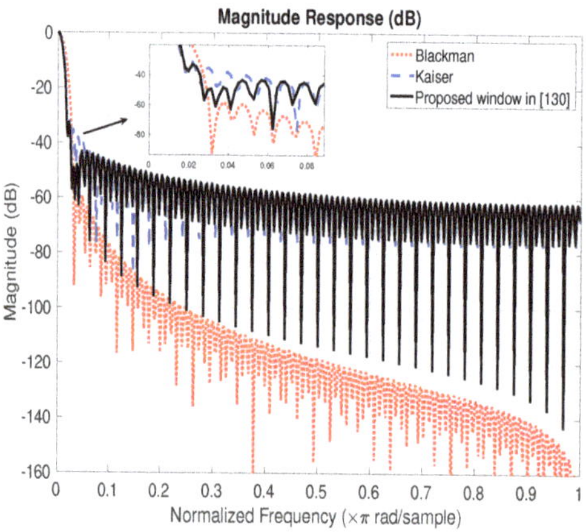

Figure 6. Frequency responses of Blackman window, Kaiser window and the new window by the GWMT [130].

Table 3. Summary of windowing based methods.

Methods	Comments	Pros and Cons
Jain's method [125]	Connecting Hamming window and Blackman window by a parameter λ.	**Pros:** Best first side-lobe level and spectrum efficiency compared to Hamming and Blackman windows. **Cons:** Large transition bandwidth.
Kumar's method [126]	Product of Hamming window and Gaussian window.	**Pros:** Excellent performance of stopband attenuation compared to Hamming and Gaussian windows. **Cons:** Under the same parameter setting, the width of main-lobe will increase.
Mottaghi-Kashtiban's method [127]	Design a four-semester raised cosine window by optimizing window parameters.	**Pros:** Narrower main-lobe under similar conditions compared to Hamming window. **Cons:** Although with improved performance of stopband attenuation, the performance gain is inconspicuous.
Rakshit's method [129]	A window function combining tangent hyperbolic function and weighted cosine series using an adjustable parameter γ.	**Pros:** Higher side-lobe roll-off ratio under the same main-lobe width compared to Gaussian window. **Cons:** Need to constantly adjust the parameters γ.
Martin-Martin's method [130]	The parameters of a four-term generalized cosine window are optimized on the basis of a given objective function.	**Pros:** Best performance of signal-to-overall-interference ratio compared to Kaiser and Blackman windows. **Cons:** High algorithm complexity.

3.3. Optimization Based Methods

In optimization based methods, some design parameters are usually given, including the filter length, the passband cutoff frequency, the stopband cutoff frequency, the number of channels, the maximum allowable distortion, and so forth. According to corresponding evaluation criteria, different objective functions and algorithms are proposed to optimize the filter parameters. This type of filter design methods is more flexible due to adjustable objective functions and constraints. Both constrained and unconstrained optimization problems can be formulated for the filter design.

3.3.1. Ababneh's Method

In Reference [131], the authors propose designing the FIR filter by the particle swarm optimization (PSO) algorithm. For each particle i, the position $\mathbf{P}_i$ and the velocity $\mathbf{V}_i$ are shown as

$$\mathbf{P}_i = (p_{i,1}, p_{i,2}, \cdots, p_{i,D}), \tag{56a}$$

$$\mathbf{V}_i = (v_{i,1}, v_{i,2}, \cdots, v_{i,D}), \tag{56b}$$

$$v_{i,d+1} = \mu v_{i,d} + c_1\beta_1(pb_{i,d} - p_{i,d}) + c_2\beta_2(gb_d - p_{i,d}), \tag{56c}$$

$$p_{i,d+1} = p_{i,d} + v_{i,d+1}, \tag{56d}$$

where D is the iteration of velocity and position, μ is the weighting function, c_1 and c_2 are the acceleration constants, β_1 and β_2 are random numbers in the range $[0,1]$. In addition, the *pbest* $pb_{i,d}$ denotes the personal best of the i_{th} particle vector at the d_{th} dimension. The *gbest* gb_d denotes the group best at the d_{th} dimension. During each iteration, the position and the velocity are calculated by the Equations (56c) and (56d), respectively.

3.3.2. Luitel's Method

In Reference [132], the differential evolution particle swarm optimization (DEPSO) algorithm combined by differential evolution and PSO is proposed to design the FIR filter. In the DEPSO algorithm, the offspring is generated by the nutation of the parent, the Gaussian distribution is considered and the parent can be represented by the *gbest*. For mutation, there are four random particles chosen from the population to produce an offspring. A mutation operator is expressed as below

$$\begin{gathered} if\ (rand < RR\ or\ d == m) \\ T_{i,d} = gb_d + \delta_{2,d}, \end{gathered} \tag{57}$$

$$\delta_{2,d} = \frac{(pb_{1,d} + pb_{2,d}) + (pb_{3,d} + pb_{4,d})}{2}, \tag{58}$$

where RR is the reproduction rate, *rand* is a random number in the range of $[0, 1]$, m means a dimension which is randomly chosen, $T_{i,d}$ is the offspring and $\delta_{2,d}$ is the weighted error in different dimensions.

3.3.3. Gupta's Method

In Reference [133], the restart PSO (RPSO) is used to design the FIR filter. Compared with PSO, the RPSO algorithm depends on the two following criteria:

Criterion 1: Terminate if the fitness's standard deviation of swarm is smaller than 10^{-3}. In this case, the particles are randomly redistributed in the search space with a probability of $1/D$.

Criterion 2: Terminate if the change in fitness of the objective function is below 10^{-8} for certain generations, then the particles are restarted by calculating derivatives to *gbest*.

Figure 7a shows the frequency responses of the PSO and RPSO algorithms, and Figure 7b shows the frequency responses of the PSO and DEPSO algorithms. Note that the filters designed by the three algorithms have similar performance. Compared to the PSO algorithm, the RPSO algorithm can prevent the results from falling into a local optimal solution, while the DEPSO algorithm has better convergence performance.

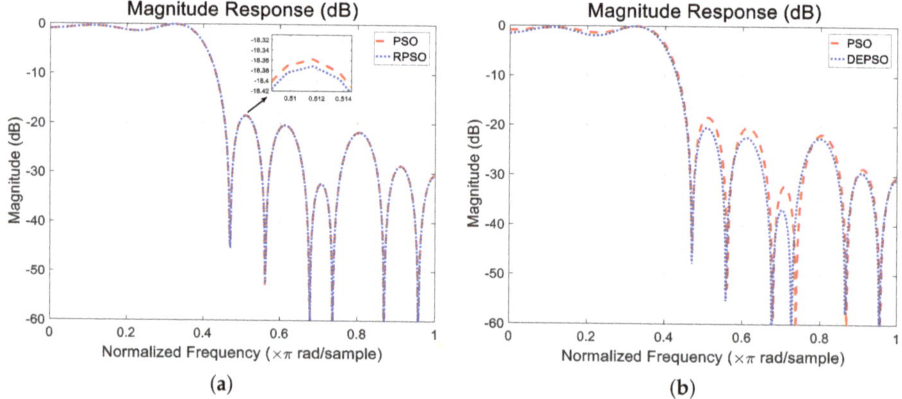

Figure 7. (**a**) Frequency responses of the designed filters by the PSO algorithm and the RPSO algorithm. (**b**) Frequency responses of the designed filters by the PSO algorithm and the DEPSO algorithm.

3.3.4. Li's Method

In Reference [134], a method to design the FIR filter based on genetic algorithm (GA) is proposed. We summarize this algorithm in Algorithm 1.

Algorithm 1 The GA Algorithm

Input: Initialize parameters: population size NP, current generation $g = 1$, maximum generation G_{max}, swarm S;

Output: The best resolution BS;

while $g \leq G_{max}$ **do**
    for $i = 1$ *to* NP **do**
        Evaluate fitness of S_g;
    end for
    for $i = 1$ *to* NP **do**
        Select operation to S_g;
    end for
    for $i = 1$ *to* $NP/2$ **do**
        Crossover operation to S_g;
    end for
    for $i = 1$ *to* NP **do**
        Mutation operation to S_g;
    end for
    for $i = 1$ *to* NP **do**
        $S_{g+1} = S_g$;
    end for
    $g = g + 1$;
end while
return The best resolution BS.

3.3.5. Karaboga's Method

In Reference [135], an FIR filter is designed based on the differential evolution (DE) algorithm. The main difference between DE algorithm and GA algorithm is that GA algorithm relies on the crossover, while DE algorithm relies on the mutation operation. The specific steps of the DE algorithm are provided in Algorithm 2.

In order to evaluate the performance of the designed FIR filters by GA and DE algorithms, the least mean squared error (LMSE) is used, and the fitness function is defined as

$$\text{Fitness} = \frac{1}{\text{LMSE}}, \tag{59}$$

and the LMSE is given as

$$\text{LMSE} = \left(\sum_k (|H_i(e^{-j\omega_k})| - |H_d(e^{-j\omega_k})|)^2 \right)^{\frac{1}{2}}, \tag{60}$$

where $H_i(e^{-j\omega_k})$ is the frequency response of ideal filter, and $H_d(e^{-j\omega_k})$ is the frequency response of designed filter.

Figure 8 shows the frequency responses of the FIR filters designed by GA and DE algorithms. The length of the two filters is $N = 21$, and Table 4 shows the control parameters of GA and DE algorithms, which have a significant effect on the performance of these two algorithms. As can be seen from Figure 8, the FIR filters designed by GA and DE algorithms have similar performance, that is, the stopband attenuation using GA is slightly better than that using DE. But the convergence speed of the DE algorithm is faster than GA.

Algorithm 2 The DE Algorithm

Input: Initialize parameters: population size NP, current generation $g = 1$, maximum generation G_{max}, dimension D, tolerance ε, swarm S, base vector s, mutant vector v, trial vector u ;
Output: The best resolution BS;
for $i = 1\ to\ NP$ **do**
  for $d = 1\ to\ D$ **do**
    $s_{i,g}^d = s_{min}^d + rand \cdot (s_{max}^d - s_{min}^d)$;
  end for
end for
while $(|f(BS)| \geq \varepsilon)\ or\ (g \leq G_{max})$ **do**
  for i=1 to NP **do**
    for d=1 to D **do**
      $v_{i,g}^d$ = Mutation($s_{i,g}^d$);
      $u_{i,g}^d$ = Crossover($s_{i,g}^d, v_{i,g}^d$);
    end for
    if $f(u_{i,G}) < f(s_{i,g})$ **then**
      $s_{i,g} = u_{i,g}$;
      if $f(s_{i,g}) < f(BS)$ **then**
        $BS = s_{i,g}$;
      end if
    else
      $s_{i,g} = s_{i,g}$;
    end if
  end for
  $g = g + 1$;
end while
return The best resolution BS.

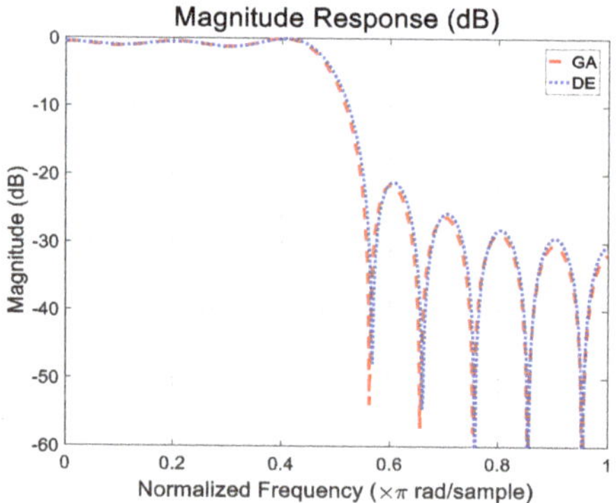

Figure 8. Frequency responses of the designed filters by the DE algorithm and the GA algorithm.

Table 4. The parameters of GA and DE algorithms.

GA Algorithm	DE Algorithm
Population size = 100	Population size = 100
Crossover rate = 0.8	Crossover rate = 0.8
Mutation rate = 0.01	Scaling factor = 0.8
Generation number = 500	Combination factor = 0.8
—	Generation number = 500

3.3.6. Chen's Method

In Reference [98], the prototype filter is designed by directly optimizing the filter coefficients. This method is mainly designed for FBMC based systems, aiming to minimize the stopband energy and constrain the ISI/ICI. The optimization problem is formulated as

$$\min_{h(0),h(1),\cdots h(N-1)} \int_{\omega_0}^{\pi} |H(e^{j\omega})|^2 d\omega$$
$$\text{s.t.} \begin{cases} h(n) = h(N-n-1), & C1 \\ \text{ISI and ICI} \leq T_1, & C2 \\ \sum_{n=0}^{N-1} (h(n))^2 = 1, & C3 \end{cases} \tag{61}$$

where T_1 is a threshold. However, this optimization problem is non-convex and non-linear, which greatly increases the computational complexity. Through a series of approximation of the constraints, the number of unknowns in the optimization problem is enormously reduced. Therefore, the optimization problem can be solved and the computational complexity becomes acceptable.

3.3.7. Hunziker's Method

In Reference [136], the design of filter banks for maximal time-frequency (TF) resolution is proposed. Two properties should be satisfied to compute the optimal DFT filter banks. Firstly, the window under the filter operation is orthogonal, thus for white input processes, the sampling at the output of the filter banks generates uncorrelated random variables. Secondly, to ensure the minimum leakage of signal components outside the target area in TF plane, the prototype window shows the best TF localization feature. To satisfy the first property, the parametrization of paraunitary filter banks in Reference [137] is used and longer pulses are extended in Reference [138]. As for the second property, the Rihaczek distribution is used in TF region, and the objective function along with the constraints is transformed into a form which can be solved by semi-definite programming (SDP) algorithm.

3.3.8. Dedeoğlu's Method

In Reference [139], the FIR filter design based on SDP is proposed. In order to satisfy both constraints on magnitude and phase responses of the filter, a non-convex quadratic constrained quadratic programming (QCQP) is constructed. Then, by relaxing the QCQP, a convex SDP is obtained, where the variables of optimization are limited to rank-1. Finally the global optimum can be found by using a convex solver. To obtain the rank-1 solution, a novel directed iterative rank refinement (DIRR) algorithm providing monotonic improvement is proposed, and a sequence of convex optimization problems are solved to minimize an adaptively chosen cost function. The performance of this design method is extensively illustrated over design cases under a variety of constraints.

3.3.9. Kobayashi's Method

A relatively new method of convex optimization is proposed to obtain the filter coefficients for FBMC, in pursuit of superior spectrum features while maintaining high symbol reconstruction quality [140]. The objective function is constructed as the minimization of out-of-band pulse energy. At the meanwhile the constraints include a maximum tolerable self-interference level and a fast spectrum decay. The main idea of the work in Reference [140] locates in transforming the formulated problem, which belongs to a non-convex QCQP, into a convex QCQP. In order to circumvent the non-convexity, a relaxation is utilized to transform the norm-2 equality constraint into a norm-1 equality constraint. Thus the transformed problem benefits from existent optimization tools. The comments, pros and cons of the above optimization based methods are summarized in Table 5.

Table 5. Summary of optimization based methods.

Methods	Comments	Pros and Cons
Ababneh's method [131]	Design the filter using the PSO algorithm.	**Pros:** Simple and intuitive. **Cons:** Poor stopband attenuation performance of the filter under low-order conditions.
Luitel's method [132]	Design the filter using the DEPSO algorithm.	**Pros:** Compared with PSO algorithm, the filter has better convergence consistency. **Cons:** Improvement of the stopband attenuation is not obvious.
Gupta's method [133]	Design the filter using the RPSO algorithm.	**Pros:** Compared with PSO algorithm, the filter has better convergence and can avoid falling into local optimum. **Cons:** Improving the performance of the stopband attenuation is not obvious.
Li's method [134]	Design the filter using the GA algorithm.	**Pros:** Obtain near global optimum solutions. **Cons:** Slow convergence rate and poor stopband attenuation performance.
Karaboga's method [135]	Design the filter using the DE algorithm.	**Pros:** Better convergence and acceptable computational complexity compared to GA. **Cons:** Improvement of the stopband attenuation is not obvious.
Chen's method [98]	Directly optimize filter coefficients.	**Pros:** Minimize the ISI/ICI and the stopband energy for FBMC modulation. **Cons:** High computational complexity.
Hunziker's method [136]	An optimization algorithm aiming at minimizing TF resolution.	**Pros:** Minimize the TF resolution. **Cons:** High first sidelobe.
Dedeoğlu's method [139]	Design the filter by convex optimization using DIRR algorithm.	**Pros:** Robust design under the phase and group delay constraints. **Cons:** Approximated solution.
Kobayashi's method [140]	Minimize the out of band pulse energy through a relaxed QCQP.	**Pros:** High symbol reconstruction performance and desirable spectral features. **Cons:** Approximated solution.

4. Discussion

Future wireless communication networks will impose strict requirements on multicarrier modulation schemes in terms of spectral efficiency, latency, computational complexity, and so forth. As shown in Table 6, crucial characteristics and applicability of five promising modulation waveforms are clearly reflected, including spectral efficiency, out of band, CP, synchronization requirement, latency, effect of frequency offset, PAPR, computational complexity, and short-burst traffic, which implicate a concrete limit to the real use of different MCM schemes. To further learn about the merits and defects of various MCM waveforms, interested readers are referred to the related works in Reference [141] for more detailed information. Since prototype filter design directly affects most characteristics of modulation waveforms such as spectral efficiency [142,143], out of band [144], it has attracted extensive research attention in the field of MCM communication. In recent years, various prototype filter design methods are designed in accordance with the above practical requirements of modulation waveforms in Table 6, and some commonly used FIR design methods for different multicarrier modulation waveforms are concluded in Table 7. The contents of this table will be continuously enriched as time goes on. In addition, we analyze the MCM system performance using three types of FIR design methods in terms of PSD and BER, which are two another important evaluation criteria [81–83,145–147]. The simulation results of PSD and BER performances are presented in Figure 9, where the FBMC is selected as the system modulation type, the number of subcarriers for FBMC is 1024, and the offset quadrature amplitude modulation (OQAM) technology is adopted. In the following, some critical characteristics and application scope of the three categories of FIR design methods are discussed and summarized, respectively.

Table 6. Respective characteristics and applicability of different multicarrier modulation waveforms.

	OFDM	FBMC	GFDM	UFMC	F-OFDM
Spectral Efficiency	Medium	High	Medium	High	Medium
Out of Band	High	Low	Low	Low	Low
Cyclic Prefix	Yes	No	Yes	No	Yes
Synchronized Requirement	High	Low	Medium	Low	Low
Latency	Short	Long	Short	Short	Short
Effect of Frequency Offset	Medium	Low	Medium	Medium	Medium
PAPR	High	High	Low	Medium	High
Computational Complexity	Low	High	High	High	Medium
Short-Burst Traffic	No	No	Yes	Yes	No

Table 7. Summary of filter design methods used in existing multicarrier modulation waveforms.

	OFDM	FBMC	GFDM	UFMC	F-OFDM
Frequency Sampling Methods	—	[51,118]	[148,149]	—	—
Windowing based Methods	[150,151]	[4]	[40,152]	[42]	[45,46,145,153]
Optimization based Methods	[154]	[140,155–157]	[158–161]	[162–164]	—

4.1. Frequency Sampling Methods

The performances of PSD and BER about different prototype filters using frequency sampling techniques are given in Figure 9a and Figure 9b, respectively. From the Figure 9a, it can be seen that the frequency sampling methods can achieve good PSD performance. The prototype filter designed by Bellanger's method has the best performance compared to other frequency sampling techniques, on the other hand, the prototype filter designed in Reference [123] has the worst BER performance while the others have the similar performance, as shown in Figure 9b.

Generally, a small number of parameters are simply designed in frequency domain, thus frequency sampling methods are inclined to optimal design and narrowband filter design where only a few non-zero values are needed. Furthermore, it is obvious that there are few side-lobes which result in better spectrum utilization and improving spectral efficiency. For time complexity, Bellanger's method achieves the order of magnitude $\mathcal{O}(N \log N)$ as most of other methods are its improved versions, and N denotes the filter length. When the value of N increases, low side-lobes may appear in the frequency domain and adjacent channel interference can be reduced, thus frequency sampling methods can be taken into consideration while low interference is required. In addition, frequency sampling techniques can satisfy the NPR condition, the research projects of 5G related networks, such as PHYDYAS and 5GNOW, consider the frequency sampling technique for FBMC modulation. Frequency sampling methods are also applied for GFDM modulation [148,149] as concluded in Table 7. However, frequency sampling methods generally have long filter length, which results in long latency, that is, there is a trade-off between high spectrum efficiency and low latency. Moreover, the position of the frequency control point is limited by N-sampling points on the frequency domain. This implies that the sampling frequency can only be an integral multiple of $2\pi/N$, resulting in that the cutoff frequency of filter is hard to control. If the cutoff frequency is freely selected, we must increase the number of sampling points N, that is, increasing the filter length, which is not conducive to short uplink burst communication in 5G scenarios.

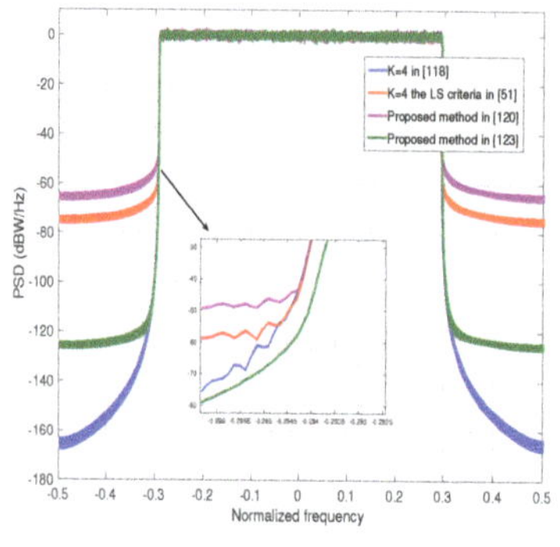

(**a**)PSDs of FBMC using frequency sampling methods.

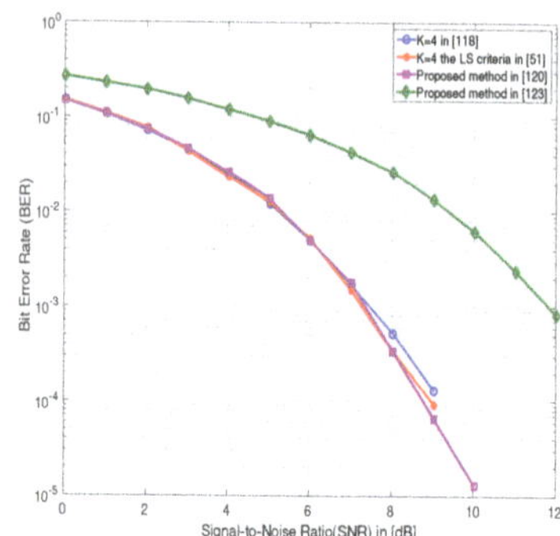

(**b**)BERs of FBMC using frequency sampling methods.

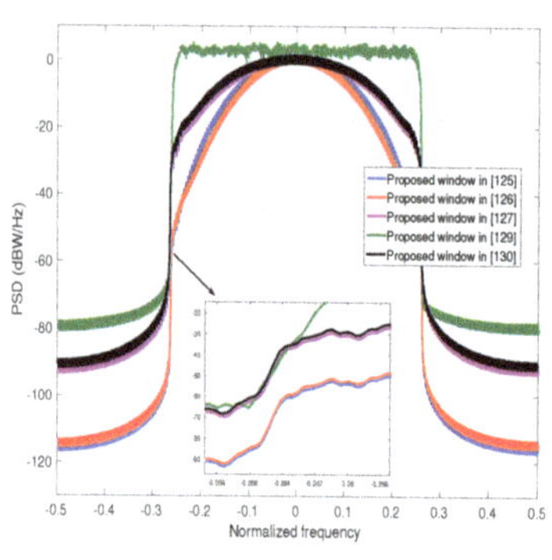

(**c**)PSDs of FBMC using windowing based methods.

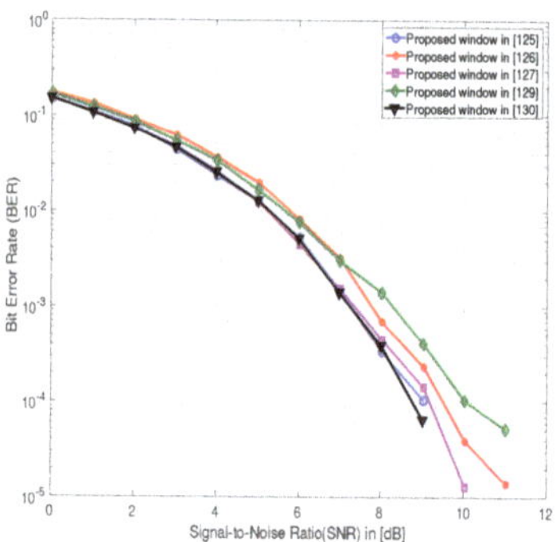

(**d**)BERs of FBMC using windowing based methods.

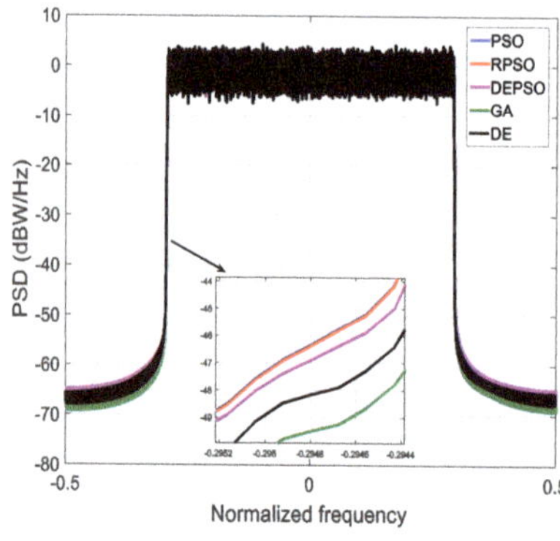

(**e**)PSDs of FBMC using optimization based methods.

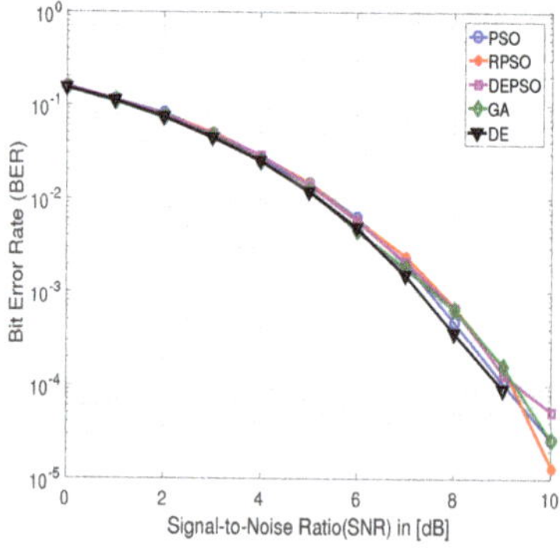

(**f**)BERs of FBMC using optimization based methods.

Figure 9. (**a**,**c**,**e**) The PSD performance comparison of FBMC based systems using different prototype filter design methods. (**b**,**d**,**f**) The BER performance comparison of FBMC based systems using different prototype filter design methods.

4.2. Windowing Based Methods

The performances of PSD and BER about different prototype filters using windowing based techniques are given in Figure 9c and Figure 9d, respectively. The aforementioned windowing based techniques have similar PSD and BER performances while the method in Reference [129] performs relatively poor compared with other methods.

The windowing based FIR filters are designed in time domain, and the basic idea is that the linear phase FIR filter multiplies by a specific windowing function, thus the time complexity is proportional to $\mathcal{O}(N)$. Generally, windowing based techniques can be seen as improved versions of classical windowing functions, nevertheless the performance improvement is not very remarkable compared with typical techniques such as Hamming window, Hanning window, Blackman window and so on. From the view of practical use, windowing based functions are easy to implement and bring little computational complexity burden on systems, therefore they are able to be used for the prototype filter design in almost all multicarrier modulation systems, as shown in Table 7. However, it is difficult to accurately control the passband cutoff frequency of the filter. The windowing process leads to a truncation effect and hence produces a transition band. Although increasing the length of the window can reduce the transition band, the amplitude of fluctuation cannot be suppressed due to Gibbs phenomenon. Changing the windowing shape can reduce the passband/stopband attenuation, but at the expense of increasing the width of transition band.

4.3. Optimization Based Methods

The performances of PSD and BER about different prototype filters using optimization based techniques are given in Figure 9e and Figure 9f, respectively. The evolutionary algorithms in References [131–135] are used for linear phase FIR filter design in recent years, and they could be taken into account in multicarrier modulation applications. In this paper, FBMC based systems using five different evolutionary algorithms are simulated, as shown in Figure 9e,f. By analyzing the simulation of the filters with the same order, it is noted that different evolutionary algorithms achieve similar PSD and BER performances.

Compared with frequency sampling techniques and windowing based techniques, the corresponding filter parameters of optimization based methods, that is more concerned with the local optimization and algorithm convergence problem, can be optimized depending on different evaluation criteria, and the selected design parameters determine the effectiveness of the solution to the optimization problem. The complexity of this type of design techniques seems to be much higher than other two techniques, for instance, the complexity of PSO based method is about the order of $LP\mathcal{O}(N^2)$, where L is the iteration size and P is the population size. This complexity is much less than QCQP based methods or non-convex optimization methods, of which the complexity is up to $\mathcal{O}(N^3)$ or higher [165]. Nevertheless, optimization based methods are much more flexible as they focus on local optimization to support the constraints imposed by various scenarios such as cognitive radio [157], massive MIMO [156] and so on. For example, when FBMC is applied for opportunistic spectrum sharing, in order to avoid interfering with other bands, well localized filters in time and frequency are prone to be designed to minimize Out of Band [140]. Similarly, there are also some optimization based methods for different waveform candidates, for example, OFDM [154], FBMC [140,155,156], GFDM [158–161], UFMC [162–164]. In general, the establishment and constraints imposed on objective functions are often highly non-linear, which increases the computational complexity and also sensitive to the selection of initial values. In addition, it is possible to fall into local optimal range and fail to identify global optimal solution, which is one of the major issues to be solved in optimization based design methods.

As discussed earlier, since multicarrier modulation can enlarge the system capacity and have the relative immunity to defend the multipath fading effect, it dominates the main stream in future communication technology. The critical part of multicarrier modulation lies in the FIR filter design. A good filter design scheme can improve some important performances not only including PSD and

BER but also some other crucial aspects, consequently increasing the competitiveness of waveform candidates and improving the communication quality to suit various application scenarios of 5G or future networks. FIR filters have been widely used in many engineering applications involved with digital signal processing, such as the communication signal [166,167], the speech signal [168,169] and the medical signal [170]. We can conclude without exaggeration that where digital signal processing is needed, where there is a shadow of FIR.

5. Conclusions

The prototype filter plays an important role in multicarrier modulation systems and the FIR filter is considered to be the suitable choice in wireless communication systems. This paper has reviewed existing FIR filter design methods which are categorized into frequency sampling methods, windowing based methods and optimization based methods. The concept and principle of each method are described in detail, and the merits and drawbacks of corresponding prototype filters are summarized. Finally, the performances of FIR design methods in different multicarrier modulation systems are evaluated and discussed in terms of PSD and BER. It is expected that this survey work can provide a basis for the selection of prototype filters in future wireless communication systems.

Author Contributions: Conceptualization, P.L. and H.Z.; methodology, H.L. and P.L.; investigation, L.J. and S.C.; writing—original draft preparation, S.C., H.L. and P.L.; writing—review and editing, L.J. and H.Z.; funding acquisition, H.Z. All authors have read and agreed to the published version of the manuscript.

Funding: This research was funded by Hubei Provincial Natural Science Foundation of China under Grant 2019CFB512 and National Natural Science Foundation of China under Grant 61501335.

Conflicts of Interest: The authors declare no conflict of interest.

References

1. Andrews, J.G.; Buzzi, S.; Choi, W.; Hanly, S.V.; Lozano, A.; Soong, A.C.K.; Zhang, J.C. What Will 5G Be? *IEEE J. Sel. Areas Commun.* **2014**, *32*, 1065–1082.
2. Agiwal, M.; Roy, A.; Saxena, N. Next Generation 5G Wireless Networks: A Comprehensive Survey. *IEEE Commun. Surv. Tutor.* **2016**, *18*, 1617–1655.
3. Gupta, A.; Jha, R.K. A Survey of 5G Network: Architecture and Emerging Technologies. *IEEE Access* **2015**, *3*, 1206–1232.
4. Sahin, A.; Guvenc, I.; Arslan, H. A Survey on Multicarrier Communications: Prototype Filters, Lattice Structures, and Implementation Aspects. *IEEE Commun. Surv. Tutor.* **2014**, *16*, 1312–1338.
5. Kaiwartya, O.; Abdullah, A.H.; Cao, Y.; Altameem, A.; Prasad, M.; Lin, C.T.; Liu, X. Internet of Vehicles: Motivation, Layered Architecture, Network Model, Challenges, and Future Aspects. *IEEE Access* **2016**, *4*, 5356–5373.
6. Chen, K.W.; Wang, C.H.; Wei, X.; Liang, Q.; Chen, C.S.; Yang, M.H.; Hung, Y.P. Vision-Based Positioning for Internet-of-Vehicles. *IEEE Trans. Intell. Transp. Syst.* **2017**, *18*, 364–376.
7. Zhang, W.; Xi, X. The innovation and development of Internet of Vehicles. *China Commun.* **2016**, *13*, 122–127.
8. Pishva, D. Internet of Things: Security and privacy issues and possible solution. In Proceedings of the International Conference Infocomm Commun. Technoligy (ICACT), Bongpyeong, South Korea, 19–22 Feburary 2017; pp. 797–808.
9. Yaqoob, I.; Ahmed, E.; Hashem, I.A.T.; Ahmed, A.I.A.; Gani, A.; Imran, M.; Guizani, M. Internet of Things Architecture: Recent Advances, Taxonomy, Requirements, and Open Challenges. *IEEE Wirel. Commun.* **2017**, *24*, 10–16.
10. Morin, E.; Maman, M.; Guizzetti, R.; Duda, A. Comparison of the Device Lifetime in Wireless Networks for the Internet of Things. *IEEE Access* **2017**, *5*, 7097–7114.
11. Rosedale, P. Virtual Reality: The Next Disruptor: A new kind of worldwide communication. *IEEE Consum. Electron. Mag.* **2017**, *6*, 48–50.
12. Serafin, S.; Erkut, C.; Kojs, J.; Nilsson, N.C.; Nordahl, R. Virtual Reality Musical Instruments: State of the Art, Design Principles, and Future Directions. *Comput. Music J.* **2016**, *40*, 22–40.

13. Douik, A.; Sorour, S.; Tembine, H.; Al-Naffouri, T.Y.; Alouini, M.S. A Game-Theoretic Framework for Network Coding Based Device-to-Device Communications. *IEEE Trans. Mob. Comput.* **2017**, *16*, 901–917.
14. Chen, B.; Yang, C.; Molisch, A.F. Cache-Enabled Device-to-Device Communications: Offloading Gain and Energy Cost. *IEEE Trans. Wirel. Commun.* **2017**, *16*, 4519–4536.
15. Haus, M.; Waqas, M.; Ding, A.Y.; Li, Y.; Tarkoma, S.; Ott, J. Security and Privacy in Device-to-Device (D2D) Communication: A Review. *IEEE Commun. Surv. Tutor.* **2017**, *19*, 1054–1079.
16. Liu, J.; Kato, N.; Ma, J.; Kadowaki, N. Device-to-Device Communication in LTE-Advanced Networks: A Survey. *IEEE Commun. Surv. Tutor.* **2015**, *17*, 1923–1940.
17. Miao, L.; Na, L.; Sun, Z. Optimal deactivated sub-carriers guard bands for spectrum pooling systems based on wavelet-based orthogonal frequency division multiplexing. *IET Commun.* **2014**, *8*, 1117–1123.
18. Boukour, T.; Chennaoui, M.; Rivenq, A.; Rouvaen, J.M.; Berbineau, M. A new WOFDM design for high data rates in the case of trains communications. In Proceedings of the Fifth IEEE International Symposium on Signal Processing and Information Technology, Athens, Greece, 21–21 December 2005; pp. 635–638.
19. Poveda, H.; Ferré, G.; Grivel, E. Way to design an orthogonal frequency-division multiple access-base station receiver disturbed by a narrowband interfering cognitive radio signal. *IET Commun.* **2015**, *9*, 1547–1554.
20. Reddy, B.S.K. Orthogonal frequency division multiple access downlink physical layer communication for IEEE 802.16-2009 standard. *IET Signal Process.* **2016**, *10*, 274–279.
21. Liu, Y.F. Complexity Analysis of Joint Subcarrier and Power Allocation for the Cellular Downlink OFDMA System. *IEEE Wirel. Commun. Lett.* **2014**, *3*, 661–664.
22. Li, Z.; Xia, X.G. An Alamouti coded OFDM transmission for cooperative systems robust to both timing errors and frequency offsets. *IEEE Trans. Wirel. Commun.* **2008**, *7*, 1839–1844.
23. Choi, K. Inter-carrier interference-free Alamouti-coded OFDM for cooperative systems with frequency offsets in non-selective fading environments. *IET Commun.* **2011**, *5*, 2125–2129.
24. Kim, B.s.; Choi, K. FADAC-OFDM: Frequency-asynchronous distributed alamouti-coded OFDM. *IEEE Trans. Veh. Technol.* **2014**, *64*, 466–480.
25. Kim, B.s.; Choi, K. Over-sampling effect in distributed Alamouti coded OFDM with frequency offset. *IET Commun.* **2016**, *10*, 2344–2351.
26. Barhumi, I.; Leus, G.; Moonen, M. Optimal training design for MIMO OFDM systems in mobile wireless channels. *IEEE Trans. Signal Process.* **2003**, *51*, 1615–1624.
27. Li, Y.G.; Winters, J.H.; Sollenberger, N.R. MIMO-OFDM for wireless communications: Signal detection with enhanced channel estimation. *IEEE Trans. Commun.* **2002**, *50*, 1471–1477.
28. Basar, E. On Multiple-Input Multiple-Output OFDM with Index Modulation for Next Generation Wireless Networks. *IEEE Trans. Signal Process.* **2016**, *64*, 3868–3878.
29. Yoshizawa, R.; Ochiai, H. Energy Efficiency Improvement of Coded OFDM Systems Based on PAPR Reduction. *IEEE Syst. J.* **2017**, *11*, 717–728.
30. Bao, H.; Fang, J.; Chen, Z.; Li, H.; Li, S. An Efficient Bayesian PAPR Reduction Method for OFDM-Based Massive MIMO Systems. *IEEE Trans. Wirel. Commun.* **2016**, *15*, 4183–4195.
31. Demir, A.F.; Elkourdi, M.; Ibrahim, M.; Arslan, H. Waveform design for 5G and beyond. *arXiv* **2019**, arXiv:1902.05999.
32. Sahin, A.; Yang, R.; Bala, E.; Beluri, M.C.; Olesen, R.L. Flexible DFT-S-OFDM: Solutions and Challenges. *IEEE Commun. Mag.* **2016**, *54*, 106–112.
33. Schellmann, M.; Zhao, Z.; Lin, H.; Siohan, P.; Rajatheva, N.; Luecken, V.; Ishaque, A. FBMC-based air interface for 5G mobile: Challenges and proposed solutions. In Proceedings of the International Conference Cognitive Radio Oriented Wireless Networks and Commun. (CROWNCOM), Oulu, Finland, 2–4 June 2014; pp. 102–107.
34. Şahin, A.; Güvenç, I.; Arslan, H. A comparative study of FBMC prototype filters in doubly dispersive channels. In Proceedings of the 2012 IEEE Globecom Workshops, Anaheim, CA, USA, 3–7 December 2012; pp. 197–203.
35. Kim, C.; Yun, Y.H.; Kim, K.; Seol, J.Y. Introduction to QAM-FBMC: From Waveform Optimization to System Design. *IEEE Commun. Mag.* **2016**, *54*, 66–73.
36. Medjahdi, Y. Interference Modeling and Performance Analysis of Asynchronous OFDM and FBMC Wireless Communication Systems. Ph.D. Thesis, Conservatoire national des arts et metiers-CNAM, Paris, France, 2012.

37. Zhang, L.; Xiao, P.; Zafar, A.; ul Quddus, A.; Tafazolli, R. FBMC system: An insight into doubly dispersive channel impact. *IEEE Trans. Veh. Technol.* **2017**, *66*, 3942–3956.
38. Farhangboroujeny, B. Filter Bank Multicarrier Modulation: A Waveform Candidate for 5G and Beyond. *Adv. Electr. Eng.* **2014**, *2014*, doi:10.1155/2014/482805.
39. Qi, Y.; Al-Imari, M. An Enabling Waveform for 5G-QAM-FBMC: Initial Analysis. *arXiv* **2018**, arXiv:1803.03169.
40. Michailow, N.; Matthé, M.; Gaspar, I.S.; Caldevilla, A.N.; Mendes, L.L.; Festag, A.; Fettweis, G. Generalized Frequency Division Multiplexing for 5th Generation Cellular Networks. *IEEE Trans. Commun.* **2014**, *62*, 3045–3061.
41. Fettweis, G.; Krondorf, M.; Bittner, S. GFDM—Generalized Frequency Division Multiplexing. In Proceedings of the IEEE Vehicle Technology Conference, Barcelona, Spain, 26-29 April 2009; pp. 1–4.
42. Vakilian, V.; Wild, T.; Schaich, F.; ten Brink, S.; Frigon, J.F. Universal-filtered multi-carrier technique for wireless systems beyond LTE. In Proceedings of the 2013 IEEE Globecom Workshops, Atlanta, GA, USA, 9–13 December 2013; pp. 223–228.
43. Wild, T.; Schaich, F.; Chen, Y. 5G air interface design based on Universal Filtered (UF-)OFDM. In Proceedings of the International Conference Digital Signal Process, Hong Kong, China, 20–23 August 2014; pp. 699–704.
44. Wen, J.; Hua, J.; Lu, W.; Zhang, Y.; Wang, D. Design of Waveform Shaping Filter in the UFMC System. *IEEE Access* **2018**, *6*, 32300–32309.
45. Abdoli, J.; Jia, M.; Ma, J. Filtered OFDM: A new waveform for future wireless systems. In Proceedings of the 2015 IEEE 16th International Workshop on Signal Processing Advances in Wireless Communications, Stockholm, Sweden, 28 June-1 July 2015; pp. 66–70.
46. Zhang, X.; Jia, M.; Chen, L.; Ma, J.; Qiu, J. Filtered-OFDM—Enabler for Flexible Waveform in the 5th Generation Cellular Networks. In Proceedings of the IEEE Global Telecommunication Conference (GLOBECOM), San Diego, CA, USA, 6–10 December 2015; pp. 1–6.
47. Yli-Kaakinen, J.; Levanen, T.; Palin, A.; Renfors, M.; Valkama, M. Generalized Fast-Convolution-based Filtered-OFDM: Techniques and Application to 5G New Radio. *arXiv* **2019**, arXiv:1903.02333.
48. Yazar, A.; Arslan, H. Selection of Waveform Parameters Using Machine Learning for 5G and Beyond. *arXiv* **2019**, arXiv:1906.03909.
49. Shu, F.; Qin, Y.; Chen, R.; Xu, L.; Shen, T.; Wan, S.; Jin, S.; Wang, J.; You, X. Directional Modulation: A Secure Solution to 5G and Beyond Mobile Networks. *arXiv* **2019**, arXiv:1803.09938.
50. Viholainen, A.; Bellanger, M.; Huchard, M. *FP7-ICT PHYDYAS-Physical Layer for Dynamic Access and Cognitive Radio*; Project Deliverable. Available online: http://www.ict-phydyas.org/delivrables/PHYDYAS-D5-1.pdf (accessed on 31 December 2010).
51. Viholainen, A.; Ihalainen, T.; Stitz, T.H.; Renfors, M. Prototype filter design for filter bank based multicarrier transmission. In Proceedings of the Signal Processing Conference, European, Glasgow, UK, 24–28 August 2009; pp. 1359–1363.
52. Konopacki, J.; Moscinska, K. A Simplified Method for IIR Filter Design with Quasi-Equiripple Passband and Least-Squares Stopband. In Proceedings of the IEEE International Conference Electronics, Circuits and Systems, Marrakech, Morocco, 11–14 December 2007; pp. 302–305.
53. Proakis, J.G. *Digital Signal Processing: Principles Algorithms and Applications*; Pearson Education India: New Delhi, India, 2001.
54. Shaeen, K.; Elizabeth, E. Prototype Filter Design Approaches for Near Perfect Reconstruction Cosine Modulated Filter Banks—A Review. *J. Signal Process. Syst.* **2015**, *81*, 183–195.
55. Chu, P. Quadrature mirror filter design for an arbitrary number of equal bandwidth channels. *IEEE Trans. Acoust. Speech Signal Process.* **1985**, *33*, 203–218.
56. Nguyen, T.Q. Near-perfect-reconstruction pseudo-QMF banks. *IEEE Trans. Signal Process.* **1994**, *42*, 65–76.
57. Saramäki, T. A generalized class of cosine-modulated filter banks. In Proceedings of the TICSP Workshop on transforms and filter banks, Tampere, Finland, 23–25 Feburary 1998; pp. 336–365.
58. Lu, W.S.; Saramaki, T.; Bregovic, R. Design of practically perfect-reconstruction cosine-modulated filter banks: A second-order cone programming approach. *IEEE Trans. Circuits Syst. I Reg. Pap.* **2004**, *51*, 552–563.
59. Kha, H.H.; Tuan, H.D.; Nguyen, T.Q. Efficient Design of Cosine-Modulated Filter Banks via Convex Optimization. *IEEE Trans. Signal Process.* **2009**, *57*, 966–976.

60. Yin, S.S.; Chan, S.C.; Tsui, K.M. On the Design of Nearly-PR and PR FIR Cosine Modulated Filter Banks Having Approximate Cosine-Rolloff Transition Band. *IEEE Trans. Circuits Syst. II Exp. Briefs 2* **2008**, *55*, 571–575.
61. Lobo, M.S.; Vandenberghe, L.; Boyd, S.; Lebret, H. Applications of second-order cone programming. *Linear Algebra Its Appl.* **1998**, *284*, 193–228.
62. Furtado, M.B.; Diniz, P.S.R.; Netto, S.L. Numerically efficient optimal design of cosine-modulated filter banks with peak-constrained least-squares behavior. *IEEE Trans. Circuits Syst. I Reg. Pap.* **2005**, *52*, 597–608.
63. Johnston, J. A filter family designed for use in quadrature mirror filter banks. In Proceedings of the ICASSP '80. IEEE International Conference Acoustics, Speech, and Signal Processing, Denver, Colorado, USA, 9–11 Aprial 1980; Volume 5, pp. 291–294.
64. Koilpillai, R.D.; Vaidyanathan, P.P. A spectral factorization approach to pseudo-QMF design. In Proceedings of the IEEE Intertational Sympoisum Circuits and Systems, Singapore, 11-14 June 1991; Volume 1, pp. 160–163.
65. Creusere, C.D.; Mitra, S.K. A simple method for designing high-quality prototype filters for M-band pseudo QMF banks. *IEEE Trans. Signal Process.* **1995**, *43*, 1005–1007.
66. Bergen, S.W.A. Design of prototype filters for cosine-modulated filter banks using the constrained least-squares method. In Proceedings of the 2005 IEEE Pacific Rim Conference Communications, Computers and Signal Processing, Victoria, BC, Canada, 24–26 August 2005; pp. 554–557.
67. Bergen, S.W.A.; Antoniou, A. An efficient closed-form design method for cosine-modulated filter banks using window functions. *Signal Process.* **2007**, *87*, 811–823.
68. Cruz-Roldan, F.; Martin-Martin, P.; Saez-Landete, J.; Blanco-Velasco, M.; Saramaki, T. A Fast Windowing-Based Technique Exploiting Spline Functions for Designing Modulated Filter Banks. *IEEE Trans. Circuits Syst. I Reg. Pap.* **2009**, *56*, 168–178.
69. Kumar, A.; Singh, G.K.; Anand, R.S. A Simple Iterative Technique for the Design of Cosine Modulated Pseudo QMF Banks; In Proceedings of ACM International Conference on Advances in Computing, Communication and Control (ICAC3-2009): Mumbai, India, 23–24 January 2009; pp. 591–596.
70. Kumar, A.; Singh, G.K.; Anand, R.S. An improved closed form design method for the cosine modulated filter banks using windowing technique. *Appl. Soft Comput.* **2011**, *11*, 3209–3217.
71. Dhabal, S.; Venkateswaran, P. Efficient Cosine Modulated Filter Bank Using Multiplierless Masking Filter and Representation of Prototype Filter Coefficients Using CSD. *Int. J. Image Graph. Signal Process.* **2012**, *4*, doi:10.5815/ijigsp.2012.10.04.
72. Chang, D.C.; Lee, D.L. Prototype Filter Design for a Cosine-Modulated Filterbank Transmultiplexer. In Proceedings of the APCCAS 2006—2006 IEEE Asia Pacific Conference Circuits and Systems, Singapore, 4–7 December 2006; pp. 454–457.
73. Soni, R.K.; Jain, A.; Saxena, R. Design of M-Band NPR Cosine-Modulated Filterbank Using IFIR Technique. *J. Signal Inf. Process.* **2010**, *1*, 35–43.
74. Diniz, P.S.R.; Barcellos, L.C.R.; Netto, S.L. Design of cosine-modulated filter bank prototype filters using the frequency-response masking approach. In Proceedings of the 2001 IEEE International Conference Acoustics, Speech, and Signal Processing, Salt Lake City, UT, USA, 7–11 May 2001; Volume 6, pp. 3621–3624.
75. Jr, M.B.F.; Diniz, P.S.R.; Netto, S.L. Optimized Prototype Filter Based on the FRM Approach for Cosine-Modulated Filter Banks. *Circuits Syst. Signal Process.* **2003**, *22*, 193–210.
76. Rosenbaum, L.; Löwenborg, P.; Johansson, M. Cosine and sine modulated FIR filter banks utilizing the frequency-response masking approach. In Proceedings of the 2003 International Symposium on Circuits and Systems, Bangkok, Thailand, 25–28 May 2003; Volume 3, pp. 882–885.
77. Rosenbaum, L.; Löwenborg, P.; Johansson, H. An Approach for Synthesis of Modulated M-Channel FIR Filter Banks Utilizing the Frequency-Response Masking Technique, EURASIP Journal on Advances in Signal Processing, **2007**; *2007*, pp. 1–13.
78. Xu, X. Design of cosine-modulated filter banks with large number of channels based on FRM technique. In Proceedings of 2011 International Conference Computer Science and Network Technology, Harbin, China, 24–26 December 2011; Volume 2, pp. 776–780.
79. Shaeen, K.; Elias, E. Design of multiplier-less cosine modulated filter banks with sharp transition using evolutionary algorithms. *Int. J. Comput. Appl.* **2013**, *68*, 1–9.

80. Bansal, B.N.; Singh, A.; Bhullar, J.S. A Review of FIR Filter Designs. In *Networking Communication and Data Knowledge Engineering*; Perez, G.M., Mishra, K.K., Tiwari, S., Trivedi, M.C., Eds.; Springer: Singapore, 2018; pp. 125–140.
81. Medjahdi, Y.; Terré, M.; le Ruyet, D.; Roviras, D. On the accuracy of PSD-based interference modeling of asynchronous OFDM/FBMC in spectrum coexistence context. In Proceedings of the 2014 11th Intertational Symposium Wireless Communication Systtem (ISWCS), Barcelona, Spain, 26–29 August 2014; pp. 638–642.
82. Zakaria, R.; Ruyet, D.L. Theoretical Analysis of the Power Spectral Density for FFT-FBMC Signals. *IEEE Commun. Lett.* **2016**, *20*, 1748–1751.
83. Bouhadda, H.; Shaiek, H.; Roviras, D.; Zayani, R.; Medjahdi, Y.; Bouallegue, R. Theoretical analysis of BER performance of nonlinearly amplified FBMC/OQAM and OFDM signals. *EURASIP J. Adv. Sign. Process.* **2014**, *2014*, 1–16.
84. Medjahdi, Y.; Terre, M.; Ruyet, D.L.; Roviras, D.; Dziri, A. The Impact of Timing Synchronization Errors on the Performance of OFDM/FBMC Systems. In Proceedings of the 2011 IEEE International Conference on Communication (ICC), Kyoto, Japan, 5–9 June 2011; pp. 1–5.
85. Cai, Y.; Qin, Z.; Cui, F.; Li, G.Y.; McCann, J.A. Modulation and Multiple Access for 5G Networks. *IEEE Commun. Surv. Tutor.* **2018**, *20*, 629–646.
86. Bala, E.; Li, J.; Yang, R. Shaping spectral leakage: A novel low-complexity transceiver architecture for cognitive radio. *IEEE Veh. Technol. Mag.* **2013**, *8*, 38–46.
87. Ijaz, A.; Zhang, L.; Xiao, P.; Tafazolli, R. Analysis of Candidate Waveforms for 5G Cellular Systems. In *Towards 5G Wireless Networks—A Physical Layer Perspective*; IntechOpen: London, UK, 2016.
88. Li, J.; Kearney, K.; Bala, E.; Yang, R. A resource block based filtered OFDM scheme and performance comparison. In Proceedings of the ICT 2013, Casablanca, Morocco, 6–8 May 2013; pp. 1–5.
89. Liu, Y.; Chen, X.; Zhong, Z.; Ai, B.; Miao, D.; Zhao, Z.; Sun, J.; Teng, Y.; Guan, H. Waveform Design for 5G Networks: Analysis and Comparison. *IEEE Access* **2017**, *5*, 19282–19292.
90. Zayani, R.; Medjahdi, Y.; Shaiek, H.; Roviras, D. WOLA-OFDM: A potential candidate for asynchronous 5G. In Proceedings of the 2016 IEEE Globecom Workshops (GC Wkshps), Washington, DC, USA, 4–8 December 2016; pp. 1–5.
91. An, C.; Ryu, H.; Li, Y.; Asharif, M.R. NW-OFDM using cyclic postfix and windowing for the eMBB waveform of the 5G/B5G mobile system. In Proceedings of the 2017 International Conference on Information and Communication Technology Convergence (ICTC), Jeju, South Korea, 18–20 October 2017; pp. 110–113.
92. Vaidyanathan, P.P. Multirate digital filters, filter banks, polyphase networks, and applications: A tutorial. *Proc. IEEE* **1990**, *78*, 56–93.
93. Tzannes, M.A.; Tzannes, M.C.; Proakis, J.; Heller, P.N. DMT systems, DWMT systems and digital filter banks. In Proceedings of ICC/SUPERCOMM'94-1994 International Conference on Communications, New Orleans, LA, USA, 1–5 May 1994; pp. 311–315.
94. Zhang, H.; Ruyet, D.L.; Terre, M. Spectral efficiency comparison between OFDM/OQAM- and OFDM-based CR networks. *Wirel. Commun. Mob. Comput.* **2009**, *9*, 1487–1501.
95. Zhang, H.; Le Ruyet, D.; Roviras, D.; Medjahdi, Y.; Sun, H. Spectral efficiency comparison of OFDM/FBMC for uplink cognitive radio networks. *EURASIP J. Adv. Signal Process.* **2010**, *2010*, 4.
96. Zhang, H.; Lv, H.; Li, P. Spectral Efficiency Analysis of Filter Bank Multi-Carrier (FBMC)-Based 5G Networks with Estimated Channel State Information (CSI). In *Towards 5G Wireless Networks—A Physical Layer Perspective*; IntechOpen: London, UK, 2016.
97. Kang, A.; Vig, R. Computational complexity analysis of FBMC-OQAM under different strategic conditions. In Proceedings of the 2014 Recent Advances in Engineering and Computational Sciences (RAECS), Chandigarh, India, 6–8 March 2014; pp. 1–6.
98. Chen, D.; Qu, D.; Jiang, T.; He, Y. Prototype Filter Optimization to Minimize Stopband Energy With NPR Constraint for Filter Bank Multicarrier Modulation Systems. *IEEE Trans. Signal Process.* **2013**, *61*, 159–169.
99. Renfors, M.; Yli-Kaakinen, J.; Harris, F.J. Analysis and Design of Efficient and Flexible Fast-Convolution Based Multirate Filter Banks. *IEEE Trans. Signal Process.* **2014**, *62*, 3768–3783.
100. Lv, S.; Zhao, J.; Ni, S. Prototype Filter Design Using Genetic Algorithm for Orientated Sidelobe Energy Suppression in FBMC. In Proceedings of the 2019 11th International Conference on Wireless Communications and Signal Processing (WCSP), Xi'an, China, 23–25 October 2019; pp. 1–6.

101. Liu, Y.; Chen, X.; Zhao, Y. Prototype Filter Design Based on Channel Estimation for FBMC/OQAM Systems. *Math. Probl. Eng.* **2019**, *2019*.
102. Rahimi, S.; Champagne, B. Joint Channel and Frequency Offset Estimation for Oversampled Perfect Reconstruction Filter Bank Transceivers. *IEEE Trans. Commun.* **2014**, *62*, 2009–2021.
103. Han, H.; Park, H. CFO estimation for QAM-FBMC systems considering non-orthogonal prototype filters. In Proceedings of the 2017 IEEE 28th Annual International Symposium on Personal, Indoor, and Mobile Radio Communications (PIMRC), Montreal, QC, Canada, 8–13 October 2017; pp. 1–5.
104. Pérez-Neira, A.I.; Caus, M.; Zakaria, R.; Ruyet, D.L.; Kofidis, E.; Haardt, M.; Mestre, X.; Cheng, Y. MIMO Signal Processing in Offset-QAM Based Filter Bank Multicarrier Systems. *IEEE Trans. Signal Process.* **2016**, *64*, 5733–5762.
105. Hu, S.; Liu, Z.; Guan, Y.L.; Jin, C.; Huang, Y.; Wu, J.M. Training Sequence Design for Efficient Channel Estimation in MIMO-FBMC Systems. *IEEE Access* **2017**, *5*, 4747–4758.
106. Mestre, X.; Gregoratti, D. Parallelized Structures for MIMO FBMC Under Strong Channel Frequency Selectivity. *IEEE Trans. Signal Process.* **2016**, *64*, 1200–1215.
107. FP7 European Project 211887 PHYDYAS (Physical Layer for Dynamic Access and Cognitive Radio). Available online: http://www.ict-phydyas.org (accessed on 1 January 2008).
108. FP7 European Project 317669 METIS (Mobile and Wireless Communications Enablers for the Twenty-Twenty Information Society) 2012. Available online: https://metis2020.com (accessed on 1 November 2012).
109. FP7 European Project 318555 5G NOW (5th Generation Non-Orthogonal Waveforms for Asynchronous Signalling) 2012. Available online: http://www.5gnow.eu (accessed on 1 September 2012).
110. Ihalainen, T.; Viholainen, A.; Stitz, T.H.; Renfors, M.; Bellanger, M. Filter Bank Based Multi-mode Multiple Access Scheme for Wireless Uplink. In Proceedings of the 2009 17th European Signal Processing Conference, Glasgow, UK, 24–28 August 2009; pp. 1354–1358.
111. Na, D.; Choi, K. Low PAPR FBMC. *IEEE Trans. Wirel. Commun.* **2017**, *17*, 182–193.
112. Choi, K. Alamouti coding for DFT spreading-based low PAPR FBMC. *IEEE Trans. Wirel. Commun.* **2018**, *18*, 926–941.
113. Liu, Z.; Xiao, P.; Hu, S. Low-PAPR Preamble Design for FBMC Systems. *IEEE Trans. Veh. Technol.* **2019**, *68*, 7869–7876.
114. Hsieh, J.H.; Tang, M.F.; Lin, M.C.; Su, B. The effect of carrier frequency offsets on an IDMA-UFMC system. In Proceedings of the 2017 Eighth International Workshop on Signal Design and Its Applications in Communications (IWSDA), Sapporo, Japan, 24–28 September 2017; pp. 89–93.
115. Zhang, L.; Ijaz, A.; Xiao, P.; Wang, K.; Qiao, D.; Imran, M.A. Optimal filter length and zero padding length design for Universal Filtered Multi-carrier (UFMC) system. *IEEE Access* **2019**, *7*, 21687–21701.
116. Padmavathi,T. ; Udayasree, P. ; Kusumakumari, C. ; Madhu,R. Performance of Universal Filter Multi Carrier in the presence of Carrier Frequency Offset. In Proceedings of the 2017 International Conference on Wireless Communications, Signal Processing and Networking (WiSPNET), Chennai, India, 22–24 March 2017; pp. 329–333.
117. Proakis, J.G.; Manolakis, D.G. *Digital Signal Processing-Principles, Algorithms and Applications*; Macmillan: New York, NY, USA, 1992.
118. Bellanger, M.G. Specification and design of a prototype filter for filter bank based multicarrier transmission. In Proceedings of the IEEE Intertional Conference Acoustics, Speech, and Signal Processing, Salt Lake City, UT, USA, 7–11 May 2001; Volume 4, pp. 2417–2420.
119. Martin, K.W. Small side-lobe filter design for multitone data-communication applications. *IEEE Trans. Circuits Syst. II Analog Digit. Signal Process.* **1998**, *45*, 1155–1161.
120. Cruz-Roldán, F.; Blanco-Velasco, M.; Martín-Martín, P.; Godino-Llorente, J.I.; Santamaría, I.; Bravo, A.M. Frequency sampling design of arbitrary-length filters for filter banks and Discrete Subband Multitone transceivers. In Proceedings of the European Signal Processing Conference, Antalya, Turkey, 4–8 September 2005; pp. 1–4.
121. Cruz-Roldan, F.; Santamaria, I.; Bravo, A.M. Frequency sampling design of prototype filters for nearly perfect reconstruction cosine-modulated filter banks. *IEEE Signal Process. Lett.* **2004**, *11*, 397–400.
122. Lin, Y.P.; Vaidyanathan, P.P. A Kaiser Window Approach For The Design Of Prototype Filters Of Cosine Modulated Filterbanks. *IEEE Signal Process. Lett.* **1998**, *5*, 132–134.

123. Cruz-Roldán, F.; Heneghan, C.; Saez-Landete, J.B.; Blanco-Velasco, M.; Amo-Lopez, P. Multi-objective optimisation technique to design digital filters for modulated multi-rate systems. *Electron. Lett.* **2008**, *44*, 827–828.
124. Salcedo-Sanz, S.; Cruz-Roldan, F.; Heneghan, C.; Yao, X. Evolutionary Design of Digital Filters With Application to Subband Coding and Data Transmission. *IEEE Trans. Signal Process.* **2007**, *55*, 1193–1203.
125. Jain, M.; Mandloi, A.S.; Parihar, A.; Shrivastava, R. A new spectral efficient window for designing of efficient FIR filter. In Proceedings of the International Conference Computer, Communication and Control (IC4), Indore, India, 10–12 September 2015; pp. 1–4.
126. Kumar, V.; Bangar, S.; Kumar, S.N.; Jit, S. Design of effective window function for FIR filters. In Proceedings of the International Conference Advances in Engineering Technology Research (ICAETR-2014), Unnao, India, 1–2 August 2014; pp. 1–5.
127. Mottaghi-Kashtiban, M.; Shayesteh, M.G. A new window function for signal spectrum analysis and FIR filter design. In Proceedings of the Iranian Conference on Electrical Engineering, Isfahan, Iran, 11–13 May 2010; pp. 215–219.
128. Harris, F.J. On the use of windows for harmonic analysis with the discrete Fourier transform. *Proc. IEEE* **1978**, *66*, 51–83.
129. Rakshit, H.; Ullah, M.A. An adjustable novel window function with its application to FIR filter design. In Proceedings of the International Conference Computer and Information Engineering (ICCIE), Rajshahi, Bangladesh, 26-27 November 2015; pp. 36–41.
130. Martin-Martin, P.; Bregovic, R.; Martin-Marcos, A.; Cruz-Roldan, F.; Saramaki, T. A Generalized Window Approach for Designing Transmultiplexers. *IEEE Trans. Circuits Syst. I Reg. Pap.* **2008**, *55*, 2696–2706.
131. Ababneh, J.I.; Bataineh, M.H. Linear phase FIR filter design using particle swarm optimization and genetic algorithms. *Digit. Signal Process.* **2008**, *18*, 657–668.
132. Luitel, B.; Venayagamoorthy, G.K. Differential evolution particle swarm optimization for digital filter design. In Proceedings of the IEEE Congress on Evolutionary Computation (IEEE World Congress on Computational Intelligence), Hong Kong, China, 1–6 June 2008; pp. 3954–3961.
133. Gupta, H.; Mandloi, D.; Jat, A.; Gupta, A.; Ojha, P. Design of linear phase low pass FIR filter using restart particle swarm optimization. In Proceedings of the Conference Advances in Computing, Communications and Informatics (ICACCI), Jaipur, India, 21–24 September 2016; pp. 2658–2661.
134. Li, K.; Liu, Y. The FIR window function design based on evolutionary algorithm. In Proceedings of the International Conference Mechatronic Science, Electric Engineering and Computer (MEC), Jilin, China,19–22 August 2011; pp. 1797–1800.
135. Karaboga, N.; Cetinkaya, B. Design of Digital FIR Filters Using Differential Evolution Algorithm. *Circuits Syst. Signal Process.* **2006**, *25*, 649–660.
136. Hunziker, T.; Rehman, U.U.; Dahlhaus, D. Spectrum sensing in cognitive radios: Design of DFT filter banks achieving maximal time-frequency resolution. In Proceedings of the 2011 8th International Conference on Information, Communications & Signal Processing, Singapore, 13–16 December 2011; pp. 1–5.
137. Cvetkovic, Z.; Vetterli, M. Tight Weyl-Heisenberg frames in l2(Z). *IEEE Trans. Signal Process.* **1998**, *46*, 1256–1259.
138. Ju, Z.; Hunziker, T.; Dahlhaus, D. Optimized Paraunitary Filter Banks for Time-Frequency Channel Diagonalization. *EURASIP J. Adv. Sig. Proc.* **2010**, *2010*, 1–12.
139. Dedeoğlu, M.; Alp, Y.K.; Arıkan, O. FIR Filter Design by Convex Optimization Using Directed Iterative Rank Refinement Algorithm. *IEEE Trans. Signal Process.* **2016**, *64*, 2209–2219.
140. Kobayashi, R.T.; Abrão, T. FBMC Prototype Filter Design via Convex Optimization. *IEEE Trans. Veh. Technol.* **2019**, *68*, 393–404.
141. Bizaki, H.K. *Towards 5G Wireless Networks-A Physical Layer Perspective*; IntechOpen: London, UK, 2016.
142. Zhang, H.; Ruyet, D.L.; Terre, M. Spectral Efficiency Analysis in OFDM and OFDM/OQAM Based Cognitive Radio Networks. In Proceedings of the VTC Spring 2009-IEEE 69th Vehicular Technology Conference, Barcelona, Spain, 26–29 April 2009; pp. 1–5.
143. Medjahdi, Y.; Terré, M.; Ruyet, D.L.; Roviras, D. On spectral efficiency of asynchronous OFDM/FBMC based cellular networks. In Proceedings of the 2011 IEEE 22nd International Symposium on Personal, Indoor and Mobile Radio Communications, Toronto, ON, Canada, 11–14 September 2011; pp. 1381–1385.

144. Kumar, R.; Tyagi, A. Computationally Efficient Mask-Compliant Spectral Precoder for OFDM Cognitive Radio. *IEEE Trans. Cogn. Commun. Netw.* **2016**, *2*, 15–23.
145. Cheng, X.; He, Y.; Ge, B.; He, C. A Filtered OFDM Using FIR Filter Based on Window Function Method. In Proceedings of the 2016 IEEE 83rd Vehicular Technology Conference (VTC Spring), Nanjing, China, 15–18 May 2016; pp. 1–5.
146. Zhong, Z.; Guo, J. Bit error rate analysis of a MIMO-generalized frequency division multiplexing scheme for 5th generation cellular systems. In Proceedings of the 2016 IEEE International Conference on Electronic Information and Communication Technology, Harbin, China, 20–22 August 2016; pp. 62–68.
147. Kumar, P.P.; Kishore, K.K. BER and PAPR Analysis of UFMC for 5G Communications. *Indian J. Sci. Technol.* **2016**, *9(S1), doi: 10.17485/.*
148. Yoshizawa, A.; Kimura, R.; Sawai, R. A Singularity-Free GFDM Modulation Scheme with Parametric Shaping Filter Sampling. In Proceedings of the 2016 IEEE 84th Vehicular Technology Conference (VTC-Fall), Montreal, QC, Canada, 18–21 September 2016; pp. 1–5.
149. Lin, D.W.; Wang, P.S. On the Configuration-Dependent Singularity of GFDM Pulse-Shaping Filter Banks. *IEEE Commun. Lett.* **2016**, *20*, 1975–1978.
150. Gudmundson, M.; Anderson, P.O. Adjacent channel interference in an OFDM system. In Proceedings of the Vehicular Technology Conference-VTC, Atlanta, GA, USA, 28 April–1 May 1996; Volume 2, pp. 918–922.
151. Pauli, M.; Kuchenbecker, P. On the reduction of the out-of-band radiation of OFDM-signals. In Proceedings of the 1998 IEEE International Conference on Communications, Atlanta, GA, USA, 7–11 June1998; Volume 3, pp. 1304–1308.
152. Xia, X.G. A family of pulse-shaping filters with ISI-free matched and unmatched filter properties. *IEEE Trans. Commun.* **1997**, *45*, 1157–1158.
153. Wu, D.; Zhang, X.; Qiu, J.; Gu, L.; Saito, Y.; Benjebbour, A.; Kishiyama, Y. A Field Trial of f-OFDM toward 5G. In Proceedings of the 2016 IEEE Globecom Workshops (GC Wkshps), Washington, DC, USA, 4–8 December 2016; pp. 1–6.
154. Vahlin, A.; Holte, N. Optimal finite duration pulses for OFDM. In Proceedings of the 1994 IEEE GLOBECOM. Commun.: The Global Bridge, San Francisco, CA, USA, 28 November–2 December 1994; Volume 1, pp. 258–262.
155. Floch, B.L.; Alard, M.; Berrou, C. Coded orthogonal frequency division multiplex [TV broadcasting]. *Proc. IEEE* **1995**, *83*, 982–996.
156. Aminjavaheri, A.; Farhang, A.; Doyle, L.E.; Farhang-Boroujeny, B. Prototype filter design for FBMC in massive MIMO channels. In Proceedings of the 2017 IEEE Intternational Conference Communication (ICC), Paris, France, 21–25 May 2017; pp. 1–6.
157. Chen, D.; Qu, D.; Jiang, T. Novel prototype filter design for FBMC based cognitive radio systems through direct optimization of filter coefficients. In Proceedings of the International Conference on Wireless Communications & Signal Processing , Suzhou, China, 21–23 October 2010; pp. 1–6.
158. Jahani-Nezhad, T.; Taban, M.R.; Tabataba, F.S. CFO estimation in GFDM systems using extended Kalman filter. In Proceedings of the 2017 Iranian Conference Electrical Engineering (ICEE), Tehran, Iran, 2–4 May 2017; pp. 1815–1819.
159. Chen, P.C.; Su, B.; Huang, Y. Matrix Characterization for GFDM: Low Complexity MMSE Receivers and Optimal Filters. *IEEE Trans. Signal Process.* **2017**, *65*, 4940–4955.
160. Nimr, A.; Matthé, M.; Zhang, D.; Fettweis, G. Optimal Radix-2 FFT Compatible Filters for GFDM. *IEEE Commun. Lett.* **2017**, *21*, 1497–1500.
161. Han, S.; Sung, Y.; Lee, Y.H. Filter Design for Generalized Frequency-Division Multiplexing. *IEEE Trans. Signal Process.* **2017**, *65*, 1644–1659.
162. Mukherjee, M.; Shu, L.; Kumar, V.; Kumar, P.; Matam, R. Reduced out-of-band radiation-based filter optimization for UFMC systems in 5G. In Proceedings of the 2015 International Wireless Communications and Mobile Computing Conference, Dubrovnik, Croatia, 24–28 August 2015; pp. 1150–1155.
163. Mukherjee, M.; Kumar, V.; Kumar, G.; Kumar, P.; Matam, R. Reduced out-of-band radiation-based filter optimization for UFMC systems in 5G—Withdrawn. In Proceedings of the 2015 International Wireless Communications and Mobile Computing Conference , Dubrovnik, Croatia, 24–28 August 2015; pp. 1–5.

164. Wang, X.; Wild, T.; Schaich, F.; dos Santos, A.F. Universal Filtered Multi-Carrier with Leakage-Based Filter Optimization. In Proceedings of the 20th European Wireless Conference, Barcelona, Spain, 14–16 May 2014; pp. 1–5.
165. Freund, R.M. *Introduction to Semidefinite Programming (SDP)*; Massachusetts Institute of Technology: Cambridge, MA, USA, 2004; p. 380.
166. Li, C.; Zhao, H.; Wu, F.; Tang, Y. Digital Self-Interference Cancellation With Variable Fractional Delay FIR Filter for Full-Duplex Radios. *IEEE Commun. Lett.* **2018**, *22*, 1082–1085.
167. Yang, A.; Yang, X.; Han, Y.; Guo, Y.; Liu, J.; Zhang, H. Wireless Channel Optimization of Internet of Things. *IEEE Access* **2018**, *6*, 54064–54074.
168. Srinivas, K.S.S.; Prahallad, K. An FIR Implementation of Zero Frequency Filtering of Speech Signals. *IEEE Trans. Audio Speech Lang. Process.* **2012**, *20*, 2613–2617.
169. Moritz, N.; Anemüller, J.; Kollmeier, B. An Auditory Inspired Amplitude Modulation Filter Bank for Robust Feature Extraction in Automatic Speech Recognition. *IEEE/ACM Trans. Audio Speech Lang. Process.* **2015**, *23*, 1926–1937.
170. Hong, Y.; Lian, Y. A Memristor-Based Continuous-Time Digital FIR Filter for Biomedical Signal Processing. *IEEE Trans. Circuits Syst. I Regul. Pap.* **2015**, *62*, 1392–1401.

MDPI
St. Alban-Anlage 66
4052 Basel
Switzerland
Tel. +41 61 683 77 34
Fax +41 61 302 89 18
www.mdpi.com

Electronics Editorial Office
E-mail: electronics@mdpi.com
www.mdpi.com/journal/electronics

www.ingramcontent.com/pod-product-compliance
Lightning Source LLC
LaVergne TN
LVHW070617170726
843515LV00004B/106

* 9 7 8 3 0 3 6 5 3 1 7 5 5 *